METABOLIC CALCULATIONS— SIMPLIFIED

DAVID P. SWAIN, PhD, FACSM

Associate Professor & Director of Wellness Institute and Research Center
ESPER Department
Old Dominion University
Norfolk, Virginia

BRIAN C. LEUTHOLTZ, PhD, FACSM

Assistant Professor
ESPER Department
Old Dominion University
Norfolk, Virginia

METABOLIC CALCULATIONS— SIMPLIFIED

Williams & Wilkins
A WAVERLY COMPANY

BALTIMORE • PHILADELPHIA • LONDON • PARIS • BANGKOK
HONG KONG • MUNICH • SYDNEY • TOKYO • WROCLAW

Editor: Eric P. Johnson
Managing Editor: Victoria Rybicki Vaughn
Marketing Manager: Christine Kushner
Production Coordinator: Raymond E. Reter
Project Editor: Jennifer D. Weir
Designer: Cotter Visual Communications, Landenberg, Pennsylvania
Cover Designer: Cotter Visual Communications, Landenberg, Pennsylvania
Typesetter: Bi-Comp, Inc., York, Pennsylvania
Printer & Binder: Mack Printing Group, Ephrata, Pennsylvania
Digitized Illustrations: Bi-Comp, Inc., York, Pennsylvania

Accurate indications, adverse reactions, and dosage schedules for drugs are provided in this book, but it is possible that they may change. The reader is urged to review the package information data of the manufacturers of the medications mentioned.

Printed in the United States of America

ISBN 0-683-30137-3

The publishers have made every effort to trace the copyright holders for borrowed material. If they have inadvertently overlooked any, they will be pleased to make the necessary arrangements at the first opportunity.

To purchase additional copies of this book, call our customer service department at **(800) 638-0672** or fax orders to **(800) 447-8438.** For other book services, including chapter reprints and large quantity sales, ask for the Special Sales department.

Canadian customers should call **(800) 665-1148,** or fax **(800) 665-0103.** For all other calls originating outside of the United States, please call **(410) 528-4223** or fax us at **(410) 528-8550.**

Visit Williams & Wilkins on the Internet: **http://www.wwilkins.com** or contact our customer service department at **custserv@wwilkins.com.** Williams & Wilkins customer service representatives are available from 8:30 am to 6:00 pm, EST, Monday through Friday, for telephone access.

98 99 00 01
2 3 4 5 6 7 8 9 10

PREFACE

This book is designed for students in the field of exercise science, candidates for American College of Sports Medicine (ACSM) certifications, and fitness professionals from personal trainers to cardiac rehabilitation specialists. The book provides simplified versions of the ACSM's metabolic equations and a very simple, straightforward approach to solving them. The equations are used to calculate the caloric expenditure or oxygen consumption of walking, running, stationary cycling, and bench stepping. Although the equations have been simplified, we must emphasize that **the simplifications have been done without altering the underlying content of the equations or the answers they yield.**

Additional chapters describe how to apply these equations to other pieces of exercise equipment, how to calculate heart rate prescriptions, and how to estimate maximal oxygen consumption from exercise tests.

Once you have read the chapters and worked through the sample problems, you will find that the equations are very easy to use. The result will be correct answers on tests and certification examinations. More importantly, we hope that your mastery of the equations will make you want to use them in your fitness practice.

David P. Swain, PhD, FACSM
Brian C. Leutholtz, PhD, FACSM

January 1997

TABLE OF CONTENTS

INTRODUCTION

Metabolic Equations: Original and Simplified

The metabolic equations contained in the "ACSM's Guidelines for Exercise Testing and Prescription" allow one to calculate oxygen consumption (VO_2) during walking, running, stationary cycling, and bench stepping. These equations are very important to fitness professionals in monitoring the exercise performance of clients and patients as well as in designing exercise prescriptions.

Although the ACSM equations are useful for these purposes, we have found that the format for the equations presented in the fifth edition of the "ACSM's Guidelines for Exercise Testing and Prescription," is sometimes difficult for students to use. For example, walking and running equations use the term meters per minute ($m \cdot min^{-1}$) for speed. Because individuals working in the United States use miles per hour (mph) for speed and those in most other countries use kilometers per hour (kph), the fitness professional must first convert the speed from the commonly used units to $m \cdot min^{-1}$ before being able to use the equation. It would be much simpler if the equation itself expressed speed in terms familiar to the user, such as mph.

Mathematical purists might cringe at the notion of mixing English units in one part of an equation with metric units in another (VO_2 is expressed in $ml \cdot min^{-1} \cdot kg^{-1}$). But our task here is not to be mathematical purists, it is to get the right answer, and simplifying the math reduces mistakes. Furthermore, for the mathematical purists out there, we should recognize that the metabolic equations contained in the ACSM's Guidelines differ from the Systeme International d'Unites (SI). For example, SI units for speed are meters per second ($m \cdot s^{-1}$), as compared to meters per minute; the units for power are Newton·meters per second ($N \cdot m \cdot s^{-1}$), or watts (W), as compared to kilogram·meters per minute ($kg \cdot m \cdot min^{-1}$).

Scientists use the metric system daily and recognize its advantages. However, our society uses the English system of units in virtually all endeavors, including fitness. The goal of this book is to make the ACSM equations user friendly. As described in Appendix I, this has been done by changing the units for speed, grade, height, etc., into commonly used terms and then adjusting the conversion factors to compensate. Furthermore, when more than one conversion factor is used by the ACSM in the same term, these are combined to eliminate a mathematical step. Finally, all the simplified equations yield answers in the same oxygen consumption (VO_2) units, that is $ml \cdot min^{-1} \cdot kg^{-1}$. The accompanying Table 1 presents both the original version of the ACSM equations and the simplified ones.

Find the VO$_2$ and Find the Workload Questions

Basically, there are two times when you might use any of these metabolic equations. The first is when a client is exercising on a given piece of equipment, and you want to be able to tell the client how many calories he or she is burning or at what level of oxygen consumption he or she is working. We will call this a find the VO_2 question. The second use of the equations is when you are writing an exercise prescription. In such a case, you want a client to exercise at a known level of energy expenditure (such as a certain percentage of his or her maximal oxygen consumption or a certain rate of calorie burning) and need to determine at what intensity on a treadmill, or other piece of equipment, you need the client to exercise. We will call this a find the workload question. Find the workload questions are more complicated than find the VO_2 questions. However, in this book, both types of questions will be solved using a simple, straight-

---

TABLE 1.

ACSM Metabolic Equations

Original
Walking
  $ml\cdot min^{-1}\cdot kg^{-1} = 3.5\ ml\cdot min^{-1}\cdot kg^{-1} + m/min \times 0.1 + grade\ (frac) \times m/min \times 1.8$
Running (treadmill)
  $ml\cdot min^{-1}\cdot kg^{-1} = 3.5\ ml\cdot min^{-1}\cdot kg^{-1} + m/min \times 0.2 + grade\ (frac) \times m/min \times 0.9$
Running (outdoors)
  $ml\cdot min^{-1}\cdot kg^{-1} = 3.5\ ml\cdot min^{-1}\cdot kg^{-1} + m/min \times 0.2 + grade\ (frac) \times m/min \times 1.8$
Leg ergometry
  $ml/min = 3.5\ ml\cdot min^{-1}\cdot kg^{-1} \times kg\ BW + None + kgm/min \times 2$
Arm ergometry
  $ml/min = 3.5\ ml\cdot min^{-1}\cdot kg^{-1} \times kg\ BW + None + kgm/min \times 3$
Bench stepping
  $ml\cdot min^{-1}\cdot kg^{-1} = (incl.\ in\ horiz.\ and\ vert.) + steps/min \times 0.35 + m/step \times steps/min \times$
                                                                          1.33×1.8

Simplified[a]
Walking
  $VO_2 = 3.5 + 2.68(speed) + 0.48(speed)(\%\ grade)$
                mph   mph
Running (treadmill)
  $VO_2 = 3.5 + 5.36(speed) + 0.24(speed)(\%\ grade)$
                mph   mph
Running (outdoors)
  $VO_2 = 3.5 + 5.36(speed) + 0.48(speed)(\%\ grade)$
                mph   mph
Leg ergometry
  $VO_2 = 3.5 + 2(workload)/BW$
                $kg\cdot m\cdot min^{-1}$ kg
Arm ergometry
  $VO_2 = 3.5 + 3(workload)/BW$
                $kg\cdot m\cdot min^{-1}$ kg
Bench stepping
  $VO_2 = 0.35(rate) + 0.061(rate)(height)$
             $steps\cdot min^{-1}$ inches

---

[a]For all simplified equations, VO_2 is in $ml\cdot min^{-1}\cdot kg^{-1}$.

---

forward approach using basic algebra. The algebraic steps will be explained as you work through the sample problems in the text.

VO$_2$: Gross Versus Net, Absolute Versus Relative

Oxygen consumption is the amount of oxygen consumed by the body in a given period, usually per minute. There are a variety of ways to express this VO_2, depending on what concept the user wishes to emphasize. The gross VO_2 is the total amount being used. During exercise, it includes both the amount the person needs for resting metabolism and the additional amount needed to perform the exercise. The net VO_2 is the amount needed for the exercise only, i.e., net VO_2 = gross VO_2 − resting VO_2. The ACSM always refers to exercise VO_2 in gross terms. Under certain circumstances, such as calculating caloric expenditure for weight loss, it might be more appropriate to use net VO_2. However, we will retain the convention of using gross VO_2 in this book to match current ACSM practice.

When VO_2 is measured in a laboratory, it is initially expressed in liters of oxygen consumed per minute (i.e. $L \cdot min^{-1}$). These are referred to as absolute units. When comparing the VO_2 of different individuals, it is often useful to express the VO_2 relative to the body mass of each person. To do this, the liters are first converted to milliliters ($1\ L = 1000\ ml$), and then this value is divided by the body mass in kilograms. The resulting relative VO_2 is in the units $ml \cdot min^{-1} \cdot kg^{-1}$. Relative units are typically used in comparing the VO_2max of different clients. A 100-kg client with a $4\ L \cdot min^{-1}\ VO_2max$ has more absolute power than a 60-kg client with a $3\ L \cdot min^{-1}\ VO_2max$. However, when expressed in relative terms, the smaller client is more "fit":

$$4000\ ml \cdot min^{-1} / 100\ kg = 40\ ml \cdot min^{-1} \cdot kg^{-1}$$
$$3000\ ml \cdot min^{-1} / 60\ kg = 50\ ml \cdot min^{-1} \cdot kg^{-1}$$

Relative units are generally used when the exercise is body mass dependent, such as walking, running, and stepping. For example, when walking on flat ground at 3 mph, persons of different sizes have the same VO_2, if it is expressed in relative units ($11.5\ ml \cdot min^{-1} \cdot kg^{-1}$), but very different absolute VO_2s. Stationary cycling is not body mass dependent, so absolute VO_2 is often used. In this book, relative VO_2 will be used for cycling because it makes it easier to write exercise prescriptions.

VO_2 may be expressed relative to other variables in addition to body mass. For example, in some settings it is useful to compare the VO_2 of different subjects based on their body surface area using the units $ml \cdot min^{-1} \cdot m^{-2}$. Throughout this book, relative VO_2 will mean the VO_2 expressed relative to body mass.

ACSM Certification

Although this book will greatly assist readers in preparing for the mathematical aspect of ACSM certification examinations, it should be noted that some sections of the book are supplemental, covering material that is useful to practitioners but that goes beyond the scope of ACSM guidelines. The supplemental material is Chapter 5 and much of Chapter 7. The explanations of the Astrand bike test and the ACSM/YMCA bike test in Chapter 7 are within ACSM guidelines. Most of Chapter 6 is within ACSM guidelines, but the specific table used to convert between percentage HRmax and percentage VO_2max is supplemental.

One concern that some readers may have is that most ACSM certifications provide the ACSM versions of the equations on the test. If you use the simplified versions, you have to memorize them for the test. It is the authors' opinion, based on experience with hundreds of certification candidates, that **it is easier to memorize and use the simplified equations than it is to be handed and then use the original ones.**

CHAPTER 1

USING THE WALKING EQUATION

WELLNESS
INSTITUTE
AND
RESEARCH
CENTER

ramili

> ## WALKING
>
> ## $VO_2 = 3.5 + 2.68(speed) + 0.48(speed)(\% \; grade)$
> $ml \cdot min^{-1} \cdot kg^{-1}$ mph   mph

As with any of the metabolic equations, you can use the walking equation to find the VO_2 or to find the workload. Finding the VO_2 is done when your client is walking, on a treadmill or outside, and you wish to know his or her oxygen consumption. If desired, this information can then be used to calculate the number of calories the client is burning. Finding the workload is done when you are designing an exercise prescription for your client, and you need to calculate a walking speed and/or treadmill grade to achieve a target intensity level.

When dealing with find the VO_2 or find the workload questions, our approach will always be the same. First, write down the appropriate metabolic equation, filling in any known values. Second, solve for the unknown.

Find the VO_2

Let's illustrate with a simple find the VO_2 problem for walking on a treadmill.

Problem 1

A client is walking on a treadmill at 3 mph up a 5% grade. What is his VO_2 in $ml \cdot min^{-1} \cdot kg^{-1}$?

$VO_2 = 3.5 + 2.68(speed) + 0.48(speed) \; (\% \; grade)$

First write down the walking equation, then fill in the known values. Remember that speed is entered in mph, and grade is entered in units of %.

$VO_2 = 3.5 + 2.68(3) + 0.48(3)(5)$

Now use basic algebra procedures to solve for the unknown. In this case, the only unknown is VO_2, which is already isolated on the left side of the equals sign. All we have to do is to perform the mathematical functions on the right. Remembering basic algebra, if terms are being multiplied together, we must perform the multiplication before they can be added to other terms. Thus, 2.68×3 is 8.04, and $0.48 \times 3 \times 5$ is 7.2.

$VO_2 = 3.5 + 8.04 + 7.2$

Now we can add these three terms.

$VO_2 = 18.74$

The answer is approximately 18.7 $ml \cdot min^{-1} \cdot kg^{-1}$.

A slight complication arises when you wish to know the client's energy expenditure in units other than $ml \cdot min^{-1} \cdot kg^{-1}$. You might prefer to use $L \cdot min^{-1}$ of oxygen consumption, METs, or $kcal \cdot min^{-1}$. This is not a serious problem. Simply solve the metabolic equation as before in the units of $ml \cdot min^{-1} \cdot kg^{-1}$ and then convert to the units you desire using the conversion factors provided in the following figure box.

METABOLIC CONVERSION FACTORS

$$VO_2 \text{ in } L \cdot min^{-1} = (VO_2 \text{ in } ml \cdot min^{-1} \cdot kg^{-1}) \times \frac{body\ wt\ in\ kg}{1000}$$

$$VO_2 \text{ in METs} = \frac{(VO_2 \text{ in } ml \cdot min^{-1} \cdot kg^{-1})}{3.5}$$

$$\text{energy expenditure in } kcal \cdot min^{-1} = (VO_2 \text{ in } L \cdot min^{-1}) \times 5\ kcal \cdot L^{-1}$$

For mathematical purists, you should note that the factor of 1000 in the conversion to $L \cdot min^{-1}$ is actually 1000 ml/L, and the factor of 3.5 in the conversion to METs is actually 3.5 $ml \cdot min^{-1} \cdot kg^{-1}/MET$.

Problem 2

A 73-kg woman is walking on a treadmill at 2.5 mph up a 6.5% grade. What is her rate of caloric expenditure?

$VO_2 = 3.5 + 2.68(speed) +$ $\qquad\qquad 0.48(speed)\,(\%\ grade)$	As with problem 1, first write down the walking equation, then fill in the known values.
$VO_2 = 3.5 + 2.68(2.5) + 0.48(2.5)(6.5)$	Now use basic algebra procedures to solve for the unknown VO_2.
$VO_2 = 3.5 + 6.7 + 7.8$	After multiplying, do the addition.
$VO_2 = 18.0$	

The answer is 18.0 $ml \cdot min^{-1} \cdot kg^{-1}$. Now we must determine the $kcal \cdot min^{-1}$ that the woman is using. From the box of metabolic conversion factors, we see that $kcal \cdot min^{-1}$ is related to VO_2 in $L \cdot min^{-1}$, and that the VO_2 in $L \cdot min^{-1}$ is related to $ml \cdot min^{-1} \cdot kg^{-1}$. Thus, to get the answer in $kcal \cdot min^{-1}$, we must first convert our current answer in $ml \cdot min^{-1} \cdot kg^{-1}$ to $L \cdot min^{-1}$ and then convert that to $kcal \cdot min^{-1}$.

$VO_2 \text{ in } ml \cdot min^{-1} \cdot kg^{-1} \times \dfrac{body\ wt\ in\ kg}{1000} =$ $\qquad\qquad\qquad VO_2 \text{ in } L \cdot min^{-1}$	First set up the conversion equation for $L \cdot min^{-1}$.
$18.0\ ml \cdot min^{-1} \cdot kg^{-1} \times \dfrac{73\ kg}{1000} = VO_2\ L \cdot min^{-1}$	Fill in known values and solve.
$1.314 = VO_2 \text{ in } L \cdot min^{-1}$	Keep all decimal places for now. We will round off at the final answer.
$\text{Expenditure in } kcal \cdot min^{-1} =$ $\qquad\qquad VO_2 \text{ in } L \cdot min^{-1} \times 5\ kcal \cdot L^{-1}$	Now set up the conversion for $kcal \cdot min^{-1}$
$\text{Expenditure in } kcal \cdot min^{-1} = 1.314 \times 5$	Fill in known values and solve.
$\text{Expenditure in } kcal \cdot min^{-1} = 6.57$	

The final answer is approximately 6.6 $kcal \cdot min^{-1}$.

Find the Workload

Now let's consider a "find the workload" question for which we already know the VO_2 at which we want our client to exercise and we need to calculate a walking speed and/or grade for the client's exercise prescription. Although this situation is the reverse of the two previous examples, it can be solved by the same straightforward approach used above. In the problem below, we will explain all the algebraic steps, for those readers who are unfamiliar with (or have forgotten) how to manipulate an equation.

Problem 3

Our prescription calls for the client to exercise with a VO_2 of 21 ml·min^{-1}·kg^{-1}. If she walks on a treadmill at 3.3 mph, at what % grade should we set the treadmill to achieve the desired VO_2?

$VO_2 = 3.5 + 2.68(\text{speed}) + 0.48(\text{speed}) (\% \text{ grade})$

As before, first write down the walking equation, then fill in the known values.

$21 = 3.5 + 2.68(3.3) + 0.48(3.3)(\% \text{ grade})$

The only unknown is % grade, which is on the right side of the equals sign. Unfortunately, it is attached to other terms, some by multiplication and some by addition. To isolate the % grade, we must **first remove terms that are being added to it and then remove terms that are being multiplied by it.** To simplify our task, let's perform the available multiplications before we proceed, i.e., $2.68 \times 3.3 = 8.844$, and $0.48 \times 3.3 = 1.584$.

$21 = 3.5 + 8.844 + 1.584(\% \text{ grade})$

Let's continue to simplify the right side of the equation by adding together the two numbers that are not attached to the % grade term, i.e., $3.5 + 8.844 = 12.344$.

$21 = 12.344 + 1.584(\% \text{ grade})$

We are done simplifying. Now we can remove the term that is being added to our unknown, i.e., the 12.344, by subtracting it from both sides of the equation. It drops out from the right side and appears as a negative value on the left side.

$21 - 12.344 = 1.584(\% \text{ grade})$

Now do the subtraction.

$8.656 = 1.584(\% \text{ grade})$

Now we need to isolate % grade by dividing both sides of the equation by 1.584. It drops out from the right side and appears in the denominator on the left side.

$8.656/1.584 = \% \text{ grade}$

Now do the division.

$5.4646 = \% \text{ grade}$

The answer is approximately 5.5%. That is, if our client walks at 3.3 mph up a 5.5% grade, her VO_2 will be approximately 21 ml·min^{-1}·kg^{-1}. Although there appear to be many steps when they are all explained in detail, with practice a person can perform them all on a calculator without writing anything down until the final answer. For now, you may find it easier to actually write out the linear equation and work your way down, filling in the steps as they are outlined here.

Let's do another problem, this time while solving for the speed.

Problem 4

We wish our next client to exercise at 35 ml·min⁻¹·kg⁻¹ while walking up a 10% grade on a treadmill. How fast must the treadmill be?

$VO_2 = 3.5 + 2.68(speed) + 0.48(speed)(\% \ grade)$	As before, first write down the walking equation, then fill in the known values.
$35 = 3.5 + 2.68(speed) + 0.48(speed)(10)$	Simplify terms by multiplying.
$35 = 3.5 + 2.68(speed) + 4.8(speed)$	We now have two terms that include speed, and one that does not. The next two steps could be done in any order, but here we will first subtract the 3.5 from both sides to isolate all the terms with speed on one side of the equation.
$35 - 3.5 = 2.68(speed) + 4.8(speed)$	
$31.5 = 2.68(speed) + 4.8(speed)$	Now we need to combine the speed terms, so that speed appears only once. This is done by the distributive property of algebra. When a single term is being multiplied by two separate numbers, these two numbers can be added together.
$31.5 = (2.68 + 4.8)(speed)$	
$31.5 = 7.48(speed)$	Now we can finalize the isolation of speed by dividing both sides by 7.48.
$31.5/7.48 = speed$	
$4.211 = speed$	

The answer is approximately 4.2 mph.

Now let's consider a complication that will be frequently encountered in the real world.

Problem 5

Your client needs to exercise at 16 ml·min⁻¹·kg⁻¹. At what speed and grade should you set the treadmill?

$VO_2 = 3.5 + 2.68(speed) + 0.48(speed)(\% \ grade)$	As before, first write down the walking equation, then fill in the known values.
$16 = 3.5 + 2.68(speed) + 0.48(speed)(\% \ grade)$	We have two unknowns, speed and grade, and the equation cannot be solved unless one of them can be determined from outside information. In the real world, we should pick a comfortable walking speed that our client can handle and then solve for % grade. If you pick the % grade first, you may come out with a speed that is too fast for walking (and therefore violates the walking equation's limits) or is too slow to be practical. So we look at our client, talk to him or her, and decide in this example on 2.5 mph before proceeding.

16 = 3.5 + 2.68(2.5) + 0.48(2.5)(% grade)

16 = 3.5 + 6.7 + 1.2(% grade)

16 = 10.2 + 1.2(% grade)

16 − 10.2 = 1.2(% grade)

5.8 = 1.2(% grade)

5.8/1.2 = % grade

4.833 = % grade

If we choose a speed of 2.5 mph for our client, we find that a grade of 4.8% yields the desired VO_2 of 16 ml·min^{-1}·kg^{-1}.

Special note: Although the ACSM does not ask open-ended questions such as this one on its certification examinations (because there would be many correct answers), your clients will present you with such questions every day.

Our final example in using the walking equation brings in the question of what units of VO_2 in which we wish to express the energy expenditure. Combined with a "find the workload" question, this is the most complicated problem we will face. Yet, using the approach we have outlined, it is actually quite simple to answer.

Problem 6

Your 86-kg client has a VO_2max of 3.1 L·min^{-1}. To exercise at 70% of capacity, what % grade would be needed on a treadmill if he is walking at 2.8 mph?

VO_2 = 3.5 + 2.68(speed) + 0.48(speed) (% grade)

You guessed it! We first write down the walking equation, then fill in the known values.

VO_2 = 3.5 + 2.68(2.8) + 0.48(2.8)(% grade)

As with the previous problem we have two unknowns, this time VO_2 and grade, and the equation cannot be solved unless one of them can be determined from outside information. This time we do not have to select one ourselves, because we have information about the VO_2. We know that the VO_2 should be 70% of 3.1 L·min^{-1}, or 0.70 × 3.1, which equals 2.17 L·min^{-1}. Our walking equation (as with all the equations in this book) calls for VO_2 in ml·min^{-1}·kg^{-1}. We must first convert VO_2 from L·min^{-1} to ml·min^{-1}·kg^{-1}. This is done using the metabolic conversion factor from the box above.

$$[VO_2 \text{ in ml·min}^{-1}\text{·kg}^{-1}] \times \frac{\text{body wt in kg}}{1000} = VO_2 \text{ in L·min}^{-1}$$

First set up the conversion equation for L·min^{-1}.

$$[VO_2 \text{ in ml·min}^{-1}\text{·kg}^{-1}] \times \frac{86 \text{ kg}}{1000} = 2.17 \text{ L·min}^{-1}$$

Fill in known values. Multiply both sides by 1000.

$[VO_2$ in ml·min^{-1}·kg^{-1} × 86 kg = 2.17 × 1000

$[VO_2$ in ml·min^{-1}·kg^{-1} × 86 kg = 2170   Divide both sides by 86.

$[VO_2$ in ml·min^{-1}·kg^{-1} = 2170/86

$[VO_2$ in ml·min^{-1}·kg^{-1} = 25.233   The answer is 25.2 ml·min^{-1}·kg^{-1}. Now we can return to our walking equation and enter this value for VO_2.

25.2 = 3.5 + 2.68(2.8) + 0.48(2.8)(% grade)   There is now only one unknown, and we can solve the equation by first simplifying the math and then isolating the unknown % grade.

25.2 = 3.5 + 7.504 + 1.344(% grade)

25.2 = 11.004 + 1.344(% grade)

25.2 − 11.004 = 1.344(% grade)

14.196 = 1.344(% grade)

14.196/1.344 = % grade

10.56 = % grade

The answer is approximately 10.6% grade. That is, our 86-kg client would be using 2.17 L·min^{-1} (which is 70% of his max) if he walked on a treadmill at 2.8 mph up a 10.6% grade.

Reference

1. American College of Sports Medicine. ACSM's guidelines for exercise testing and prescription. 5th ed. Baltimore: Williams & Wilkins, 1995:278–283.

Sample Problems Using the Walking Equation

1. A client walks on a treadmill at 3 mph up a 14% grade. What is her VO_2 in ml·min^{-1}·kg^{-1} and in METs?

2. A 70-kg client walks outdoors at 3.7 mph. How many calories would he burn in 30 minutes?

3. A patient with cardiac disease walks on a treadmill at 1.7 mph up a 10% grade during the first stage of the Bruce protocol. What VO_2, in ml·min^{-1}·kg^{-1}, do you expect him to be using during this stage?

4. Your 132-lb client is walking on a treadmill at 2.8 mph up a 5% grade for 40 minutes, 3 times a week. Assuming no change in her diet, how long will it take her to lose 10 pounds of fat with this exercise routine?

5. You want a client to exercise at 18 ml·min^{-1}·kg^{-1} while walking on a treadmill at 2.1 mph. What grade should be used?

6. A 62-kg client has a VO_2max of 3.0 L·min^{-1}. To have him exercise at 60% of functional capacity, what grade on a treadmill would you use if he is walking at 2.5 mph?

7. You want a 92-kg client to exercise at an intensity of 5 kcal·min^{-1} while walking outdoors. At what speed should he walk in mph and what should be his pace per mile?

8. You want a client to exercise at 7 METs on a treadmill. Your treadmill is broken and will only work at a 10% grade. At what speed should you set the treadmill?

9. Your 48-kg client is walking on a treadmill at 3 mph up a 6.5% grade. If she does this for 1 hour, what is the total amount of oxygen, in liters, that she will consume?

10. You want a client with a VO_2max of 42 ml·min^{-1}·kg^{-1} to walk outdoors at 70% of his functional capacity. What would be his speed?

Answers to the Walking Equation Sample Problems

1. Answer: 31.7 ml·min^{-1}·kg^{-1}; 9.1 METs

   Solution: This is straightforward. Set up the equation and solve for VO_2.

   $VO_2 = 3.5 + 2.68(3) + 0.48(3)(14)$

   $VO_2 = 3.5 + 8.04 + 20.16$

   $VO_2 = 31.7$

   The first part of the answer is 31.7 ml·min^{-1}·kg^{-1}. To convert this to METs, check the metabolic conversion factors and divide the VO_2 by 3.5.

   $31.7/3.5 = 9.06$ METs, or approximately 9.1 METs

2. Answer: 141 kcal

   Solution: Do not let the ''calories in 30 minutes'' throw you. This is a simple ''find the VO_2'' question with a twist at the end. Set up the equation and solve for VO_2. Note that the vertical component drops out. We assume an overall flat course, i.e., 0% grade, for the outdoor walk (unless stated otherwise).

   $VO_2 = 3.5 + 2.68(3.7) + 0.48(3.7)(0)$

   $VO_2 = 3.5 + 9.916 + 0$

   $VO_2 = 13.416$

   The client is at a VO_2 of approximately 13.4 ml·min^{-1}·kg^{-1}. We need to convert this to L·min^{-1} and then to kcal·min^{-1}.

   $13.416 \text{ ml·min}^{-1}\text{·kg}^{-1} \times \dfrac{70 \text{ kg}}{1000} = 0.939 \text{ L·min}^{-1}$

   $0.939 \text{ L·min}^{-1} \times 5 \text{ kcal·L}^{-1} = 4.695 \text{ kcal·min}^{-1}$

   Because the client is burning approximately 4.7 kcal·min^{-1} and is doing this for 30 minutes, the total number of kcal burned is:

   $4.7 \times 30 = 141$ kcal

   Note that this is the *gross* energy expenditure, i.e., the total for both the resting component (from the 3.5 ml·min^{-1}·kg^{-1} in the walking equation) and the exercise component. (If you wanted only the *net* energy expenditure, you would delete the 3.5 from the walking equation at the start.)

3. Answer: 16.2 ml·min^{-1}·kg^{-1}

   Solution:

   $VO_2 = 3.5 + 2.68(1.7) + 0.48(1.7)(10)$

   $VO_2 = 3.5 + 4.556 + 8.16$

   $VO_2 = 16.216$

   Note: the ACSM states that the walking equation is accurate for speeds between 1.9 and 3.7 mph. However, it is often used for speeds slightly beyond this range, as long as the person is walking normally.

4. Answer: 55 weeks

Solution: As in sample problem 2, we first start with the walking equation to get VO_2, then convert to kcal·min^{-1}. Then we will address the question of weight loss.

$VO_2 = 3.5 + 2.68(2.8) + 0.48(2.8)(5)$

$VO_2 = 3.5 + 7.504 + 6.72$

$VO_2 = 17.724$

To convert this VO_2 into L·min^{-1}, we need the client's weight in kg. Because there are approximately 2.2 lb per kg, her body mass is 130 lb/2.2 lb·kg^{-1} = 60 kg.

$$17.724 \text{ ml·min}^{-1}\text{·kg}^{-1} \times \frac{60 \text{ kg}}{1000} = 1.0634 \text{ L·min}^{-1}$$

1.0634 L·min^{-1} × 5 kcal·L^{-1} = 5.317 kcal·min^{-1}

If the client burns approximately 5.3 kcal·min^{-1} for 40 minutes for 3 times a week, then she burns 5.3 × 40 × 3 = 636 kcal each week. Because 1 lb of fat has a caloric value of 3500 kcal and 10 lb has 35,000 kcal, it will take her:

35,000 kcal/636 kcal·wk^{-1} = 55 weeks

It will take a little over 1 year to burn off this much fat through moderate exercise. (Note: we used *gross* energy expenditure to determine her rate of fat loss. This is not really proper because if she had chosen to rest during those 40 minutes, she still would have burned some of the calories we are counting for fat loss. It would be more accurate to use *net* energy expenditure to calculate the fat loss due to the exercise itself. However, because the ACSM uses gross values in such determinations, we did so in this example. When working with clients in the real world, you may perform this calculation with net values if you prefer.)

5. Answer: 8.8%

Solution: Here is a straightforward find the workload question. Set up the walking equation and solve for the unknown.

$18 = 3.5 + 2.68(2.1) + 0.48(2.1)(\% \text{ grade})$

$18 = 3.5 + 5.628 + 1.008(\% \text{ grade})$

$18 = 9.128 + 1.008(\% \text{ grade})$

$18 - 9.128 = 1.008(\% \text{ grade})$

$8.872 = 1.008(\% \text{ grade})$

$8.872/1.008 = \% \text{ grade}$

$8.802 = \% \text{ grade}$

6. Answer: 16.5%

Solution: If you set up your walking equation, you discover you have two unknowns, VO_2 and grade. Let's determine the VO_2 in ml·min^{-1}·kg^{-1} from the available information and then use the walking equation to solve for the unknown grade.

$0.60 \times 3.1 = 1.86$ L·min^{-1}

$$1.86 \text{ L·min}^{-1} \times \frac{1000}{62 \text{ kg}} = 30 \text{ ml·min}^{-1}\text{·kg}^{-1}$$

$30 = 3.5 + 2.68(2.5) + 0.48(2.5)(\% \text{ grade})$

$30 = 3.5 + 6.7 + 1.2(\% \text{ grade})$

$30 = 10.2 + 1.2(\% \text{ grade})$

$30 - 10.2 = 1.2(\%\ grade)$

$19.8 = 1.2(\%\ grade)$

$19.8/1.2 = \%\ grade$

$16.5 = \%\ grade$

7. Answer: 2.8 mph; 21 minutes, 26 seconds per mile

Solution: Convert the kcal·min^{-1} to a VO$_2$, then use the walking equation to solve for speed.

$5\ kcal\cdot min^{-1}/5\ kcal\cdot L^{-1} = 1.0\ L\cdot min^{-1}$

$1.0\ L\cdot min^{-1} \times \dfrac{1000}{92\ kg} = 10.87\ ml\cdot min^{-1}\cdot kg^{-1}$

$10.87 = 3.5 + 2.68(speed) + 0.48(speed)(0)$

$10.87 = 3.5 + 2.68(speed) + 0$

$10.87 - 3.5 = 2.68(speed)$

$7.47 = 2.68(speed)$

$7.47/2.68 = speed$

$2.79 = speed$

The speed is approximately 2.8 mph. What is the pace per mile, i.e., how many minutes and seconds should he plan to take to walk 1 mile? If he goes 2.8 miles in 60 minutes, then he is taking 60/2.8 = 21.43 minutes for each mile. The fractional 0.43 minutes represents 0.43 × 60 seconds, or 25.7 seconds. So his pace is about 21:26 per mile, depending on where you did your rounding off.

8. Answer: 2.8 mph

Solution: Convert the METs to ml·min^{-1}·kg^{-1} and use the walking equation to solve for the unknown speed.

$7\ METs \times 3.5 = 24.5\ ml\cdot min^{-1}\cdot kg^{-1}$

$24.5 = 3.5 + 2.68(speed) + 0.48(speed)(10)$

$24.5 - 3.5 = 2.68(speed) + 4.8(speed)$

$21 = (2.68 + 4.8)(speed)$

$21 = 7.48(speed)$

$21/7.48 = speed$

$2.807 = speed$

9. Answer: 60.2 L

Solution: Use the walking equation to solve for VO$_2$, then convert to liters.

$VO_2 = 3.5 + 2.68(3) + 0.48(3)(6.5)$

$VO_2 = 3.5 + 8.04 + 9.36$

$VO_2 = 20.9$

$20.9\ ml\cdot min^{-1}\cdot kg^{-1} \times \dfrac{48\ kg}{1000} = 1.0032\ L\cdot min^{-1}$

$1.0032\ L\cdot min^{-1} \times 60\ min = 60.192\ L$

10. Answer: ???

Solution: The exercise VO$_2$ is 70% of 42 ml·min^{-1}·kg^{-1}, or 0.70 × 42 = 29.4. Let's put that in the walking equation and see what happens when we solve for speed.

29.4 = 3.5 + 2.68(speed) + 0.48(speed)(0)

29.4 = 3.5 + 2.68(speed) + 0

29.4 − 3.5 = 2.68(speed)

25.9 = 2.68(speed)

25.9/2.68 = speed

9.664 = speed

The "answer" is approximately 9.7 mph. But can people walk that fast? No. Even if they could, you could not claim that it would yield a VO$_2$ of 29.4 ml·min^{-1}·kg^{-1}. This is because the walking equation is not accurate for walking speeds significantly greater than 3.7 mph when the person is race-walking rather than walking normally. Is the answer to run at 9.7 mph? No. You cannot use the walking equation to calculate a running speed. You must use the running equation to do that. Which brings us to the next chapter.

CHAPTER 2

USING THE RUNNING EQUATIONS

16
17 Very hard
18
19 Very, very hard
20

TREADMILL RUNNING

VO$_2$ = 3.5 + 5.36(speed) + 0.24(speed)(% grade)
ml·min^{-1}·kg^{-1} mph   mph

Treadmill

This version of the running equation is used for running on a treadmill. Later in this chapter, we will explain how to modify the equation for running outdoors.

Let's begin with a simple "find the VO$_2$" question. Remember, the steps we take in answering the question are always: 1) write down the appropriate equation, 2) fill in the known values, and 3) solve for the unknown.

Problem 1

A client is running on a treadmill at 6 mph up a 3% grade. What is his VO$_2$ in ml·min^{-1}·kg^{-1} and in METs?

VO$_2$ = 3.5 + 5.36(speed) + 0.24(speed) (% grade)	First write down the running equation, then fill in the known values. Remember that speed is entered in mph, and grade is entered in units of %.
VO$_2$ = 3.5 + 5.36(6) + 0.24(6)(3)	Now use basic algebra procedures to solve for the unknown. In this case, the only unknown is VO$_2$, which is already isolated on the left side of the equals sign. First we'll do the multiplications.
VO$_2$ = 3.5 + 32.16 + 4.32	Now we can add these three terms.
VO$_2$ = 39.98	

The answer is approximately 40.0 ml·min^{-1}·kg^{-1}. Dividing this value by 3.5 yields the answer in METs: 40.0/3.5 = 11.4 METs.

Now let's turn the problem around by asking a "find the workload" question.

Problem 2

You wish to have your client exercise at 53 ml·min^{-1}·kg^{-1} by running on a treadmill. He is comfortable running at 5 mph. At what grade should you set the treadmill?

VO$_2$ = 3.5 + 5.36(speed) + 0.24(speed) (% grade)	First write down the running equation, then fill in the known values.
53 = 3.5 + 5.36(5) + 0.24(5)(% grade)	In this case, the only unknown is % grade. Let's simplify the other terms, multiplications first.
53 = 3.5 + 26.8 + 1.2(% grade)	Now add the 3.5 to the 26.8.

53 = 30.3 + 1.2(% grade)

Now let's isolate % grade on the right side of the equation. First, by subtracting the 30.3.

53 − 30.3 = 1.2(% grade)

22.7 = 1.2(% grade)

And now, by dividing by 1.2.

22.7/1.2 = % grade

18.92 = % grade

The answer is approximately 18.9%. That is, if our client runs on a treadmill at 5 mph up an 18.9% grade, his VO_2 will be approximately 53 ml·min^{-1}·kg^{-1}.

As you can see, using the running equation is exactly the same and just as easy as using the walking equation.

Outdoors

The only complication in using the running equation comes when you try to take it outside. The walking equation is exactly the same on a treadmill or outdoors. However, as explained in Appendix I, this is not true for the running equation. **For outdoor running, you must change the vertical conversion factor from 0.24 to 0.48.**

> ## OUTDOOR RUNNING
>
> **VO_2 = 3.5 + 5.36(speed) + 0.48(speed)(% grade)**
> ml·min^{-1}·kg^{-1} mph   mph

To illustrate the use of the outdoor version of the running equation, let's redo problem 1 from this chapter but change the exercise from a treadmill to outside.

Problem 3

A client is running outside at 6 mph up a hill with a steady 3% grade. What is his VO_2 in ml·min^{-1}·kg^{-1} and in METs?

VO_2 = 3.5 + 5.36(speed) + 0.48(speed) (% grade)

First write down the outdoor version of the running equation, then fill in the known values.

VO_2 = 3.5 + 5.36(6) + 0.48(6)(3)

Now solve as before.

VO_2 = 3.5 + 32.16 + 8.64

VO_2 = 44.3

The answer is approximately 44.3 ml·min^{-1}·kg^{-1}, or 44.3/3.5 = 12.7 METs. Note that it is a little bit harder to run outdoors than on a treadmill at the same grade.

Of course, you seldom know what the grade is outside. Most of the time the jogging course is flat or nearly flat with equal uphill and downhill sections. In such cases, enter the grade as zero, and the vertical component drops out. Let's redo sample problem 10 from chapter 1, in which the walking equation gave a very high speed (9.7 mph). Because that is too fast to walk, should you just have the client run that fast? No, because the wrong equation was used to obtain that speed. Let's do it properly with the running equation.

Problem 4

You want a client with a VO_2max of 42 ml·min^{-1}·kg^{-1} to run outdoors at 70% of his functional capacity. What would be his speed?

The exercise VO_2 is $0.70 \times 42 = 29.4$ ml·min^{-1}·kg^{-1}. We put that into the running equation and solve for speed.

$VO_2 = 3.5 + 5.36(\text{speed}) + 0.48(\text{speed})(\% \text{ grade})$ First write down the outdoor running equation, then fill in the known values.

$29.4 = 3.5 + 5.36(\text{speed}) + 0.48(\text{speed})(0)$ Zero times anything is zero, so the vertical component drops out. (It really did not matter if we had used the treadmill or the outdoor version of the running equation, because it still would have been zero.)

$29.4 = 3.5 + 5.36(\text{speed}) + 0$

$29.4 = 3.5 + 5.36(\text{speed})$

$29.4 - 3.5 = 5.36(\text{speed})$

$25.9 = 5.36(\text{speed})$

$25.9/5.36 = \text{speed}$

$4.8 = \text{speed}$

The answer is approximately 4.8 mph (not the 9.7 mph the walking equation gave us). According to the ACSM, true running is at 5 mph or faster. However, the equation can be used down to 3 mph if the subject is jogging (both feet leave the ground between steps) and not just walking fast.

In case you need to use the walking and running equations when this book is not handy, memorizing them is easy. Just memorize the walking equation. Then, remember that the treadmill running equation has double the horizontal conversion factor (from 2.68 to 5.36) and half the vertical conversion factor (from 0.48 to 0.24). If running up a hill outdoors, go back to 0.48 for the vertical component.

Reference

1. American College of Sports Medicine. ACSM's guidelines for exercise testing and prescription. 5th ed. Baltimore: Williams & Wilkins, 1995:278–283.

Sample Problems Using the Running Equations

1. A client runs on a treadmill at 5.2 mph up a 7% grade. What is her VO_2 in ml·min^{-1}·kg^{-1} and in METs?

2. A 65-kg client runs outdoors at 6 mph. What is her VO_2 in L·min^{-1}?

3. Your client has a functional capacity of 12 METs. You want him to run on a treadmill at 80% of his capacity. If he runs at 4.6 mph, what grade should be used?

4. You want a 200-lb client to burn 0.5 lb of fat each week through exercise. If he runs 3 days a week at 7 mph, how long should he run each day?

5. Your 80-kg client is running on a treadmill at 8 mph up a 3% grade. What is his energy expenditure in $kcal \cdot min^{-1}$?

6. You want your client to run on a treadmill at 50 $ml \cdot min^{-1} \cdot kg^{-1}$. If the treadmill is set at a 5% grade, what speed should be used?

7. Your personal training client completed a 10K race (6.2 miles) in 32 minutes, 24 seconds. What was her VO_2 during the race in $ml \cdot min^{-1} \cdot kg^{-1}$?

8. If the racer in question 7 had to climb a 5% grade hill, how much would she have to slow down (assuming she kept her VO_2 the same)?

9. What is the VO_2 in $L \cdot min^{-1}$ of a 78-kg client while running on a treadmill at 6 mph up a 4% grade?

10. Your client has a VO_2max of 58 $ml \cdot min^{-1} \cdot kg^{-1}$. You want him to exercise at 85% of this level on a treadmill. If the speed is set at 7 mph, what % grade should be used?

Answers to the Running Equations Sample Problems

1. Answer: 40.1 $ml \cdot min^{-1} \cdot kg^{-1}$; 11.5 METs

   Solution:

   $VO_2 = 3.5 + 5.36(5.2) + 0.24(5.2)(7)$

   $VO_2 = 3.5 + 27.872 + 8.736$

   $VO_2 = 40.108$

   The first part of the answer is 40.1 $ml \cdot min^{-1} \cdot kg^{-1}$. To get METs: 40.1/3.5 = 11.5.

2. Answer: 2.3 $L \cdot min^{-1}$

   Solution:

   $VO_2 = 3.5 + 5.36(6) + 0.48(6)(0)$

   $VO_2 = 3.5 + 32.16 + 0$

   $VO_2 = 35.66$

   Now convert to $L \cdot min^{-1}$: 35.66 × 65/1000 = 2.3179. The answer is approximately 2.3 $L \cdot min^{-1}$.

3. Answer: 4.9%

   Solution: The VO_2 is 80% of 12 METs. In $ml \cdot min^{-1} \cdot kg^{-1}$, this would be: 0.80 × 12 × 3.5 = 33.6. Now solve for % grade.

   $33.6 = 3.5 + 5.36(4.6) + 0.24(4.6)(\% \text{ grade})$

   $33.6 = 3.5 + 24.656 + 1.104(\% \text{ grade})$

   $33.6 = 28.156 + 1.104(\% \text{ grade})$

   $33.6 - 28.156 = 1.104(\% \text{ grade})$

   $5.444 = 1.104(\% \text{ grade})$

   $5.444/1.104 = \% \text{ grade}$

   $4.931 = \% \text{ grade}$

4. Answer: approximately 31 minutes

   Solution: Sounds pretty complicated. Let's start by setting up the running equation and solving for VO_2. From the wording of the question, we'll assume he's running on flat ground.

   $VO_2 = 3.5 + 5.36(7) + 0$

   $VO_2 = 3.5 + 37.52$

   $VO_2 = 41.02$

   Now we have his VO_2 as 41.0 ml·min^{-1}·kg^{-1}. The question asks how long it would take to burn 0.5 lb of fat. The common denominator here is kilocalories. A half pound of fat has $3500/2 = 1750$ kcal. Running 3 times a week, he needs to burn $1750/3 = 583$ kcal per session. Let's convert his VO_2 into kilocalories per minute to see how long he needs to exercise to burn off that many calories. To convert to L·min^{-1}, we multiply by his body weight in kg ($200/2.2 = 90.9$) and divide by 1000.

   $41.0 \times 90.9/1000 = 3.73$ L·min^{-1}

   Multiply by 5 to get kilocalories per minute: $3.73 \times 5 = 18.7$. Now divide this into the number of kilocalories he needs to burn each session: $583/18.7 = 31.2$ minutes. (Remember, this is based on the gross energy expenditure. If you used net values, it would take a little longer.)

5. Answer: 20.9 kcal·min^{-1}

   Solution:

   $VO_2 = 3.5 + 5.36(8) + 0.24(8)(3)$

   $VO_2 = 3.5 + 42.88 + 5.76$

   $VO_2 = 52.14$

   To convert to kcal·min^{-1}, first convert to L·min^{-1}, then multiply that answer by 5:

   $52.14 \times 80/1000 = 4.1712$ L·min^{-1}

   $4.1712 \times 5 = 20.856$ kcal·min^{-1}

6. Answer: 7.1 mph

   Solution:

   $50.0 = 3.5 + 5.36(\text{speed}) + 0.24(\text{speed})(5)$

   $50.0 = 3.5 + 5.36(\text{speed}) + 1.2(\text{speed})$

   $50.0 = 3.5 + (5.36 + 1.2)(\text{speed})$

   $50.0 = 3.5 + 6.56(\text{speed})$

   $50.0 - 3.5 = 6.56(\text{speed})$

   $46.5 = 6.56(\text{speed})$

   $46.5/6.56 = \text{speed}$

   $7.088 = \text{speed}$

7. Answer: 65.1 ml·min^{-1}·kg^{-1}

   Solution: If you set up the equation, you will find that the speed is missing. Determine the speed from the distance (6.2 miles) and the time (32 minutes, 24 seconds) provided. The time must be converted to hours: 24 of 60 seconds per minute is $24/60 = 0.40$ minutes; 32.40 of 60 minutes per hour is $32.40/60 = 0.540$ hours. Thus, the speed is 6.2 miles/ 0.540 hours = 11.5 mph.

$VO_2 = 3.5 + 5.36(11.5) + 0$

$VO_2 = 3.5 + 61.64$

$VO_2 = 65.14$

8. Answer: from 11.5 mph to 7.9 mph

Solution: Keeping the VO_2 from the previous answer, solve for the new speed. Remember to use the outdoor running equation, in which the vertical conversion factor is 0.48.

$65.1 = 3.5 + 5.36(speed) + 0.48(speed)(5)$

$65.1 = 3.5 + 5.36(speed) + 2.4(speed)$

$65.1 = 3.5 + (5.36 + 2.4)(speed)$

$65.1 = 3.5 + 7.76(speed)$

$65.1 - 3.5 = 7.76(speed)$

$61.6 = 7.76(speed)$

$61.6/7.76 = speed$

$7.938 = speed$

What is the pace per mile for this speed? Just divide the mph into 60: $60/7.94 = 7.56$ min per mile.

The 0.56 minutes is $0.56 \times 60 = 34$ seconds. The pace must slow to 7 minutes, 34 seconds from her original 5 minutes, 14 seconds (determined from the 32.4 minutes/6.2 miles in question 7).

9. Answer: 3.2 L·min^{-1}

Solution:

$VO_2 = 3.5 + 5.36(6) + 0.24(6)(4)$

$VO_2 = 3.5 + 32.16 + 5.76$

$VO_2 = 41.42$

Convert to L·min^{-1}: $41.42 \times 78/1000 = 3.23076$

10. Answer: 4.9%

Solution:

$0.85 \times 58 = 3.5 + 5.36(7) + 0.24(7)(\% \text{ grade})$

$49.3 = 3.5 + 37.52 + 1.68(\% \text{ grade})$

$49.3 = 41.02 + 1.68(\% \text{ grade})$

$49.3 - 41.02 = 1.68(\% \text{ grade})$

$8.28 = 1.68(\% \text{ grade})$

$8.28/1.68 = \% \text{ grade}$

$4.929 = \% \text{ grade}$

CHAPTER 3

USING THE CYCLE ERGOMETRY EQUATIONS

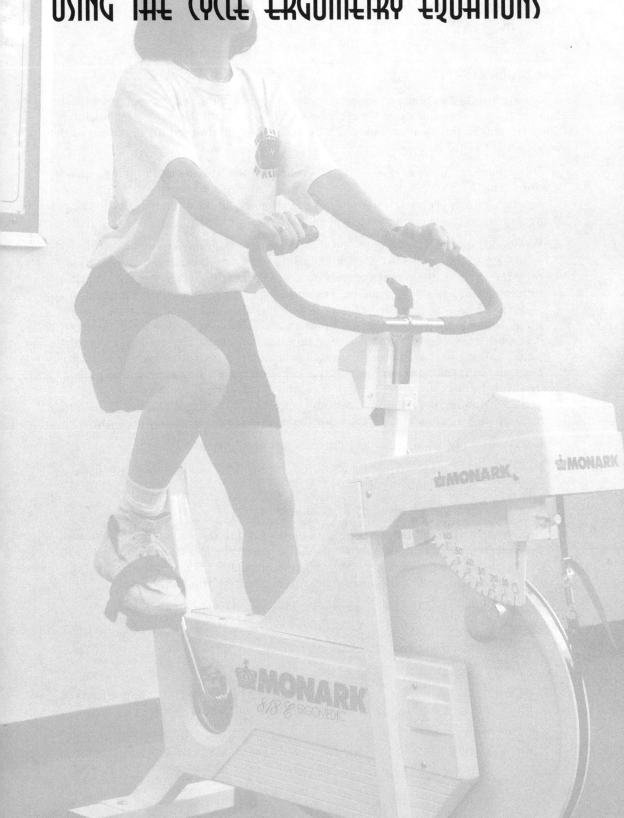

LEG ERGOMETRY

$VO_2 = 3.5 + 2(\text{workload})/BW$
$\text{ml·min}^{-1}\text{·kg}^{-1}$ kg·m·min^{-1} kg

Leg Ergometry

There are two cycle ergometry equations: one for performing cycling exercise with the legs, as shown above, and one for cycling exercise done with the arms, which will be covered later in this chapter. In both cases, the units that are used for the terms in the equation are as follows:

1. VO_2 comes out in $\text{ml·min}^{-1}\text{·kg}^{-1}$, just as it does in the walking and running equations. The ACSM version of the leg ergometry equation comes out in ml·min^{-1}. As described in Appendix I, we have rearranged the equation so that VO_2 is in the per kg format, making it easier to use in exercise prescriptions.

2. Workload, or power, is entered into the equation in the units kg·m·min^{-1} (referred to as "kgm/min" by the ACSM). When using stationary cycle ergometers such as the Monark bike, the power depends on how fast the flywheel is moving against a set amount of frictional resistance. The resistance setting is generally reported in kilograms. (This is technically incorrect. Kilograms are units of mass, not of resistive force. It would be more proper to report the resistive force in Newtons. However, it is more convenient to use kilograms, because the force scale on the bike is calibrated in kilograms.) The speed of the flywheel depends on how fast the subject is pedaling, in revolutions per minute (rpm), and on the gearing used to drive the flywheel. In the case of the Monark bike, the flywheel moves 6 meters for each pedal revolution. For Tunturi and BodyGuard bikes, the flywheel moves 3 meters for each pedal revolution. To calculate the workload, use the following formula:

   workload = (kg setting)(flywheel distance/revolution)(rpm)

   For example, if the resistance on a Monark bike is 3 kg, and the subject is pedaling at 60 rpm, the workload would be:

   $3 \text{ kg} \times 6 \text{ m·rev}^{-1} \times 60 \text{ rev·min}^{-1} = 1080 \text{ kg·m·min}^{-1}$

3. Body weight (BW) of the subject must be entered into the equation in kilograms. To convert pounds into kilograms, divide the pounds by 2.2. (Common mistake to avoid: do not use the subject's body weight in calculating the workload. The "kg" in the workload calculation comes from the resistance setting on the bike.)

Let's do a simple "find the VO_2" question to see how this equation works.

Problem 1

A 65-kg client is cycling on a stationary ergometer at 1200 kg·m·min^{-1}. What is his VO_2 in $\text{ml·min}^{-1}\text{·kg}^{-1}$?

$VO_2 = 3.5 + 2(\text{workload})/BW$	First write down the leg ergometry equation, then fill in the known values.
$VO_2 = 3.5 + 2(1200)/65$	It does not matter if we multiply the 1200 by 2 and then divide it by 65, or if we divide the 1200 by 65 and then multiply it by 2; either

way gives the same answer. Let's multiply by 2 first.

$VO_2 = 3.5 + 2400/65$ Now divide by 65.

$VO_2 = 3.5 + 36.92$ Finally, add the 3.5 to give the final answer.

$VO_2 = 40.42$

The answer is approximately 40.4 ml·min^{-1}·kg^{-1}.

Now let's try it when we have to figure out the workload ourselves.

Problem 2

A 73-kg client is cycling on a Monark bike at 50 rpm with a force setting of 2.2 kg. What is his VO$_2$ in ml·min^{-1}·kg^{-1}?

$VO_2 = 3.5 + 2(\text{workload})/BW$ First write down the leg ergometry equation, then fill in the known values.

$VO_2 = 3.5 + 2(\text{workload})/73$ We have two unknowns, VO$_2$ and workload. The equation cannot be solved until we determine workload from the bike settings we were given. Let's do that first.

workload = (kg setting)(flywheel distance/revolution)(rpm)   Write down the workload equation and fill in the values.

workload = (2.2)(6)(50)   Now solve.

workload = 660 kg·m·min^{-1} Now we can return to our leg ergometry equation and enter this value for the workload.

$VO_2 = 3.5 + 2(660)/73$ Solve as in problem 1.

$VO_2 = 3.5 + 1320/73$

$VO_2 = 3.5 + 18.08$

$VO_2 = 21.58$

The answer is approximately 21.6 ml·min^{-1}·kg^{-1}.

We are now ready for a harder "find the workload" question.

Problem 3

You want your 86-kg client to exercise at 70% of her 6 MET capacity on a BodyGuard stationary bike. What workload, in kg·m·min^{-1}, should you use? If she cycles at 50 rpm, what resistance setting is needed?

$VO_2 = 3.5 + 2(\text{workload})/BW$ First write down the leg ergometry equation, then fill in the known values.

$VO_2 = 3.5 + 2(\text{workload})/86$ We have two unknowns, VO$_2$ and workload. We can determine the VO$_2$ from the information provided in the question. Seventy percent of 6 METs is: $0.70 \times 6 = 4.2$ METs. There are 3.5 ml·min^{-1}·kg^{-1} in each MET, so: $4.2 \times 3.5 = 14.7$ ml·min^{-1}·kg^{-1}. Now let's enter that in our equation and proceed.

$14.7 = 3.5 + 2(\text{workload})/86$

We must isolate workload on the right side of the equation by removing the other terms. As we did in the walking and running equations, first subtract the 3.5, which is held more "loosely" to the unknown by addition than the other terms that are attached to the unknown by multiplication and division.

$14.7 - 3.5 = 2(\text{workload})/86$

$11.2 = 2(\text{workload})/86$

Multiplication and division are of equal "tightness," so it does not matter which one we remove first. Let's start by dividing both sides by 2.

$11.2/2 = (\text{workload})/86$

$5.6 = (\text{workload})/86$

Now multiply by 86.

$5.6 \times 86 = \text{workload}$

$481.6 = \text{workload}$

The answer is approximately 482 kg·m·min^{-1}. If the client is to accomplish this on a BodyGuard bike at 50 rpm, we need to set up the workload equation to solve for the resistance setting.

workload = (force setting)(flywheel distance/ revolution)(rpm)

Set up the equation and fill in the known values. Remember that a BodyGuard flywheel travels 3 meters for each pedal revolution.

$482 = (\text{force setting})(3)(50)$

$482 = (\text{force setting})(150)$

$482/150 = \text{force setting}$

$3.21 = \text{force setting}$

The final answer is that our client can exercise at 70% of her MET capacity by pedaling a BodyGuard bike at 50 rpm with a force setting of about 3.2 kg. This would give her a workload of 482 kg·m·min^{-1}.

We have been expressing the workload in terms of kg·m·min^{-1}. The proper units for workload, or power, are actually watts. Watts are well known units that are commonly used; thus, we need to know how to convert between them and kg·m·min^{-1}. Watts are Newton·meters per second. Because the force of gravity on 1 kg at sea level is approximately 9.8 N and there are 60 seconds in 1 minute, the relationship between watts and kg·m·min^{-1} is:

$$1\ \text{W} = \frac{1\ \text{N·m}}{1\ \text{sec}} \times \frac{1\ \text{kg}}{9.8\ \text{N}} \times \frac{60\ \text{sec}}{1\ \text{min}} = 6.12\ \text{kg·m·min}^{-1}$$

To convert watts to $kg \cdot m \cdot min^{-1}$ accurately, multiply the watts by 6.12. **However, the ACSM uses a conversion factor of 6.** Thus, when we want to convert watts to $kg \cdot m \cdot min^{-1}$, we will multiply the watts by 6; when we want to convert $kg \cdot m \cdot min^{-1}$ to watts, we will divide the $kg \cdot m \cdot min^{-1}$ by 6.

Problem 4

A 52-kg client is cycling on a stationary ergometer at 250 W. What is her VO_2 in $ml \cdot min^{-1} \cdot kg^{-1}$ and in $L \cdot min^{-1}$?

$VO_2 = 3.5 + 2(workload)/BW$

First write down the leg ergometry equation. The workload in the equation calls for $kg \cdot m \cdot min^{-1}$. Because the workload in the question is 250 W, we first multiply this by 6 to convert it to $kg \cdot m \cdot min^{-1}$: $250 \times 6 = 1500$ $kg \cdot m \cdot min^{-1}$.

$VO_2 = 3.5 + 2(1500)/52$

$VO_2 = 3.5 + 3000/52$

$VO_2 = 3.5 + 57.69$

$VO_2 = 61.2$

The answer is approximately 61.2 $ml \cdot min^{-1} \cdot kg^{-1}$. As in several previous examples, to express this in $L \cdot min^{-1}$, we multiply by the body weight and divide by 1000: $61.2 \times 52/1000 = 3.18 \ L \cdot min^{-1}$.

An interesting aspect of cycle ergometry is that the net oxygen cost (that due only to the exercise) expressed in *absolute* terms (i.e., in $L \cdot min^{-1}$) depends solely on the workload. That is, it is the same value for all subjects regardless of their size. This is exactly the opposite of the situation with walking and running, for which the workload (speed and grade) produces the same *relative* oxygen cost (i.e., in $ml \cdot min^{-1} \cdot kg^{-1}$) for all subjects. To illustrate this aspect of cycling ergometry, let's redo problem 4 with a larger client.

Problem 5

A 120-kg client is cycling on a stationary ergometer at 250 W. What is his VO_2 in $ml \cdot min^{-1} \cdot kg^{-1}$ and in $L \cdot min^{-1}$?

$VO_2 = 3.5 + 2(workload)/BW$

First write down the leg ergometry equation. As before, the 250 W is approximately equal to 1500 $kg \cdot m \cdot min^{-1}$.

$VO_2 = 3.5 + 2(1500)/120$

$VO_2 = 3.5 + 3000/120$

If we stop here, notice that the workload produces a VO_2 of 3000 $ml \cdot min^{-1}$ (or 3.0 $L \cdot min^{-1}$), just as it did for the client in problem 4. The only difference is that there is a larger body (120 kg) to divide with.

$VO_2 = 3.5 + 25.0$

$VO_2 = 28.5$

The answer is approximately 28.5 ml·min⁻¹·kg⁻¹. Multiplying by body weight and dividing by 1000 gives us: $28.5 \times 120/1000 = 3.42$ L·min⁻¹. Note that the *relative VO₂* is much less than it was for the smaller client (28.5 versus 61.2 ml·min⁻¹·kg⁻¹)—it is easier for the larger client to do the same amount of work. But the *absolute VO₂* is quite similar (3.42 versus 3.18 L·min⁻¹). The slightly greater amount of absolute VO₂ for the larger client is because our equation provides gross VO₂, resting plus exercise, and the larger client has a larger resting VO₂ in absolute terms. The net VO₂, that due only to the exercise, is 3.0 L·min⁻¹ for both subjects.

Arm Ergometry

What about arm ergometry? Cycling exercise can be performed with the arms, sometimes referred to as "arm cranking." One way to perform arm cranking is to place a Monark leg ergometer on a table and sit in front of it, grasping the pedals with the hands. Alternatively, one can use specifically designed arm ergometers, such as the Monark Rehab Trainer. To calculate the expected VO₂ during arm ergometry, we must modify the leg ergometry equation. Specifically, instead of multiplying the workload by 2, it is multiplied by 3, because the arms are less efficient than the legs.

ARM ERGOMETRY

$VO_2 = 3.5 + 3(\text{workload})/BW$
ml·min⁻¹·kg⁻¹   kg·m·min⁻¹   kg

The workload is calculated from the resistance setting, flywheel distance per revolution, and rpm, just as it was for leg ergometry. However, be aware that the flywheels on the specialized arm ergometers made by Monark travel 2.4 meters per revolution, not 6. Of course, when you crank a Monark leg ergometer with your arms, its flywheel still goes 6 meters per revolution.

Problem 6

You want the 86-kg client from problem 3 to exercise at 14.7 ml·min⁻¹·kg⁻¹ on a Monark arm ergometer. What workload, in kg·m·min⁻¹, should you use? If she cranks at 50 rpm, what force setting is needed?

VO₂ = 3.5 + 3(workload)/BW	First write down the arm ergometry equation, then fill in the known values.
14.7 = 3.5 + 3(workload)/86	Now solve for workload.
14.7 − 3.5 = 3(workload)/86	
11.2 = 3(workload)/86	
11.2/3 = (workload)/86	
3.73 = (workload)/86	
3.73 × 86 = workload	
320.8 = workload	

The answer is approximately 321 kg·m·min^{-1}. This is much less than she did with her legs in problem 3 (482 kg·m·min^{-1}) to achieve the same VO$_2$. Now let's set up the workload equation to solve for the resistance setting.

workload = (force setting)(flywheel distance/ revolution)(rpm)	Set up the equation and fill in the known values. Remember that a Monark arm ergometer flywheel travels 2.4 meters for each "pedal" revolution.

321 = (force setting)(2.4)(50)

321 = (force setting)(120)

321/120 = force setting

2.675 = force setting

Our client needs to crank the arm ergometer at 50 rpm at a force setting of approximately 2.7 kg.

There are other pieces of equipment for performing arm exercise, such as arm/leg bikes (e.g., the Schwinn AirDyne) and rowing machines. These machines are not operated solely with the arms; therefore, the arm ergometry equation is not appropriate for estimating the VO$_2$ of these activities. We will discuss how to apply the ergometry equations to these devices in Chapter 5.

Reference

1. American College of Sports Medicine. ACSM's guidelines for exercise testing and prescription. 5th ed. Baltimore: Williams & Wilkins, 1995:278–283.

Sample Problems Using the Cycle Ergometry Equations

1. A 76-kg client is exercising on a stationary bike at 1800 kg·m·min^{-1}. What is his estimated VO$_2$ in ml·min^{-1}·kg^{-1}?

2. A 62-kg client is exercising on a stationary bike at 100 W. What is her estimated VO$_2$ in ml·min^{-1}·kg^{-1} and in L·min^{-1}?

3. You want your 81-kg client to exercise at 7.5 kcal·min^{-1} on a stationary bike. What workload is needed in kg·m·min^{-1}?

4. You want your 100-kg client to exercise at 60% of his 4.5 L·min^{-1} capacity on a Monark bike. If he pedals at 60 rpm, what resistance setting is needed?

5. You want your 47-kg client to exercise at 10 METs on a Tunturi bike. If she pedals at 50 rpm, what resistance setting is needed?

6. A 68-kg client is arm cranking on a Monark leg ergometer at 900 kg·m·min^{-1}. What is his VO$_2$ in ml·min^{-1}·kg^{-1}?

7. A 91-kg, severely deconditioned patient is arm cranking on a Monark arm ergometer at 70 rpm with a resistance setting of 1 kg. What is her VO$_2$ in ml·min^{-1}·kg^{-1}?

8. A 65-kg client is pedaling a leg ergometer at 225 W. At what MET level is he exercising?

9. You want your 80-kg client to burn 2000 kcal per week on a stationary bike. If she exercises at 1200 kg·m·min^{-1}, how many minutes per week must she exercise?

10. Your 53-kg client has a VO$_2$peak of 2.1 L·min^{-1} during arm exercise. What workload in kg·m·min^{-1} would place her at 60% of her capacity?

Answers to the Cycle Ergometry Equation Sample Problems

1. Answer: $50.9 \text{ ml·min}^{-1}\text{·kg}^{-1}$

   Solution: This is a simple "find the VO_2" problem. Set up the equation and solve for the VO_2.

   $VO_2 = 3.5 + 2(1800)/76$

   $VO_2 = 3.5 + 3600/76$

   $VO_2 = 3.5 + 47.37$

   $VO_2 = 50.87$

2. Answer: $22.9 \text{ ml·min}^{-1}\text{·kg}^{-1}$; 1.42 L·min^{-1}

   Solution: Find the VO_2, but first we convert the watts to $kg·m·min^{-1}$ by multiplying the watts by 6:

   $100 \times 6 = 600 \text{ kg·m·min}^{-1}$

   $VO_2 = 3.5 + 2(600)/62$

   $VO_2 = 3.5 + 1200/62$

   $VO_2 = 3.5 + 19.35$

   $VO_2 = 22.85$

   To convert this VO_2 into $L·min^{-1}$, multiply by body weight and divide by 1000.

   $22.85 \times 62/1000 = 1.417 \text{ L·min}^{-1}$

3. Answer: approximately $608 \text{ kg·m·min}^{-1}$

   Solution: Neither the workload nor the VO_2 is known, but the VO_2 can be obtained from the energy expenditure.

   $7.5 \text{ kcal·min}^{-1}/5 \text{ kcal·L}^{-1} = 1.5 \text{ L·min}^{-1}$

   $1.5 \text{ L·min}^{-1} \times 1000/81 \text{ kg} = 18.5 \text{ ml·min}^{-1}\text{·kg}^{-1}$

   $18.5 = 3.5 + 2(\text{workload})/81$

   $18.5 - 3.5 = 2(\text{workload})/81$

   $15.0 = 2(\text{workload})/81$

   $15.0/2 = (\text{workload})/81$

   $7.5 = (\text{workload})/81$

   $7.5 \times 81 = \text{workload}$

   $607.5 = \text{workload}$

4. Answer: approximately 3.3 kg

   Solution: The VO_2 is determined from the information provided. The absolute VO_2 during exercise is 60% of 4.5 L·min^{-1}.

   $0.60 \times 4.5 = 2.7 \text{ L·min}^{-1}$

   The relative VO_2 is the absolute value times 1000, divided by body weight.

   $2.7 \times 1000/100 = 27 \text{ ml·min}^{-1}\text{·kg}^{-1}$

   $27 = 3.5 + 2(\text{workload})/100$

   $27 - 3.5 = 2(\text{workload})/100$

   $23.5 = 2(\text{workload})/100$

23.5/2 = (workload)/100

11.75 = (workload)/100

11.75 × 100 = workload

1175 = workload

The workload on the Monark bike is approximately 1175 kg·m·min^{-1}. To determine the resistance setting on the bike, set up the workload equation and solve.

1175 = (resistance setting)(6 m·rev^{-1})(60 rpm)

1175 = (resistance setting)(360)

1175/360 = resistance setting

3.26 = resistance setting

5. Answer: approximately 4.9 kg

   Solution: METs times 3.5 gives us the VO$_2$ in the proper units.

   10 × 3.5 = 35 ml·min^{-1}·kg^{-1}

   35 = 3.5 + 2(workload)/47

   35 − 3.5 = 2(workload)/47

   31.5 = 2(workload)/47

   31.5/2 = (workload)/47

   15.75 = (workload)/47

   15.75 × 47 = workload

   740.25 = workload

   The workload on the Tunturi bike is approximately 740 kg·m·min^{-1}. To determine the resistance setting on the bike, set up the workload equation and solve. Remember that the Tunturi flywheel only travels 3 meters per pedal revolution.

   740 = (resistance setting)(3 m·rev^{-1})(50 rpm)

   740 = (resistance setting)(150)

   740/150 = resistance setting

   4.93 = resistance setting

6. Answer: 43.2 ml·min^{-1}·kg^{-1}

   Solution: For arm ergometry, remember to use 3 for the workload multiplier.

   VO$_2$ = 3.5 + 3(900)/68

   VO$_2$ = 3.5 + 2700/68

   VO$_2$ = 3.5 + 39.71

   VO$_2$ = 43.21

7. Answer: 9.0 ml·min^{-1}·kg^{-1}

   Solution: This is a "find the VO$_2$" question, but we have to determine the workload from the information given. Set up the workload equation, remembering that a Monark arm ergometer's flywheel travels 2.4 meters per "pedal" revolution.

workload = (1 kg)(2.4 m·rev^{-1})(70 rpm) = 168 kg·m·min^{-1}

VO$_2$ = 3.5 + 3(168)/91

VO$_2$ = 3.5 + 504/91

VO$_2$ = 3.5 + 5.54

VO$_2$ = 9.04

8. Answer: 12.9 METs

Solution: Watts times 6 gives us the workload to enter into the leg ergometry equation: 225 × 6 = 1350 kg·m·min^{-1}.

VO$_2$ = 3.5 + 2(1350)/65

VO$_2$ = 3.5 + 2700/65

VO$_2$ = 3.5 + 41.54

VO$_2$ = 45.04

To convert to METs, we divide by 3.5: 45.04/3.5 = 12.87 METs.

9. Answer: approximately 149 minutes per week

Solution: First let's find her VO$_2$ and then convert that to caloric expenditure.

VO$_2$ = 3.5 + 2(1200)/80

VO$_2$ = 3.5 + 2400/80

VO$_2$ = 3.5 + 30

VO$_2$ = 33.5

Convert to L·min^{-1}: 33.5 × 80/1000 = 2.68 L·min^{-1}

Convert to kcal·min^{-1}: 2.68 × 5 = 13.4 kcal·min^{-1}

Because she needs to burn 2000 kcal per week and she is burning them at a rate of 13.4 each minute, the number of minutes is:

2000 kcal/13.4 kcal·min^{-1} = 149.3 minutes

10. Answer: approximately 359 kg·m·min^{-1}

Solution: The VO$_2$ is 60% of 2.1 L·min^{-1}.

0.60 × 2.1 = 1.26 L·min^{-1}

Convert to relative VO$_2$.

1.26 × 1000/53 = 23.8 ml·min^{-1}·kg^{-1}

23.8 = 3.5 + 3(workload)/53

23.8 − 3.5 = 3(workload)/53

20.3 = 3(workload)/53

20.3/3 = (workload)/53

6.77 = (workload)/53

6.77 × 53 = workload

358.8 = workload

CHAPTER 4

USING THE STEPPING EQUATION

STEPPING

$$VO_2 = 0.35(\text{rate}) + 0.061(\text{rate})(\text{height})$$
$$\text{ml·min}^{-1}\text{·kg}^{-1} \quad \text{steps·min}^{-1} \qquad\qquad \text{inches}$$

The bench stepping equation is formatted differently from the walking, running, and ergometry equations in the preceding chapters: the resting component of 3.5 ml·min^{-1}·kg^{-1} is not seen in the formula. This makes it appear that the equation gives net VO$_2$, the VO$_2$ due to exercise alone. However, this is not the case. According to the ACSM, the equation still gives gross VO$_2$, because the calculation of the "horizontal" component (i.e., 0.35 × stepping rate) includes the resting VO$_2$.

In using the simplified equation, the stepping rate is expressed in steps per minute, and the height of the bench is expressed in inches. If the reported bench height is in centimeters, convert to inches by dividing the centimeters by 2.54.

The stepping rate is based on one "step" being a complete cycle of "up-up, down-down." Beginning with both feet on the floor, the client starts by placing the left leg up on the bench (the first "up"); then the client pushes with the left leg, raising the body until the right leg is placed on the bench (the second "up"). The client then steps down, placing the left leg on the floor (the first "down"). Finally, the client places the right leg on the floor (the second "down"). Of course, the choice of which leg goes first does not matter, except to note that the first leg (the left one in the example here) does almost all of the concentric (lifting) work, and the second leg does almost all of the eccentric (lowering) work. The motor pattern for bench stepping is very different from climbing a set of stairs.

If a metronome is being used to pace the client, it should be set at a rate that is 4 times the desired stepping rate. For example, if you want the client to do 20 steps per minute, the metronome would be set at 80 beats per minute, and the client would perform one leg movement with each beat.

Let's perform one "find the VO$_2$" problem and one "find the workload" problem to illustrate the equation's use.

Problem 1

A 62-kg client is stepping on a 12-inch box at a rate of 30 steps·min^{-1}. What is her VO$_2$ in ml·min^{-1}·kg^{-1}, METs, and L·min^{-1}?

$VO_2 = 0.35(\text{rate}) + 0.061(\text{rate})(\text{height})$	First write down the stepping equation, then fill in the known values. Remember that height is entered in inches.
$VO_2 = 0.35(30) + 0.061(30)(12)$	Now use basic algebra procedures to solve for the unknown. In this case, the only unknown is VO$_2$, which is already isolated on the left side of the equals sign. First we'll do the multiplications.
$VO_2 = 10.5 + 21.96$	Now we add these two terms.
$VO_2 = 32.46$	

The answer is approximately 32.5 ml·min^{-1}·kg^{-1}. Divide this value by 3.5 to obtain METs: 32.5/3.5 = 9.3 METs. Multiply the relative VO$_2$ by the body weight and divide by 1000 to get the absolute VO$_2$: 32.5 × 62/1000 = 2.015, or approximately 2.02 L·min^{-1}.

Problem 2

You want your client to exercise at 20 ml·min^{-1}·kg^{-1} by stepping on an 8-inch bench. What stepping rate is needed, and at what rate should you set the metronome?

VO$_2$ = 0.35(rate) + 0.061(rate)(height)

First write down the stepping equation, then fill in the known values.

20 = 0.35(rate) + 0.061(rate)(8)

Simplify the math by doing the multiplication in the last term, 0.061 × 8 = 0.488.

20 = 0.35(rate) + 0.488(rate)

Two terms consist of rate multiplied by a number. According to the distributive property of algebra, we can add the two numbers together.

20 = (0.35 + 0.488)(rate)

20 = 0.838(rate)

Now isolate rate by dividing by 0.838.

20/0.838 = rate

23.87 = rate

On an 8-inch bench, the stepping rate that will elicit a VO$_2$ of 20 ml·min^{-1}·kg^{-1} is approximately 24 steps·min^{-1}. Multiply this value by 4 to get the metronome rate: 24 × 4 = 96 beats·min^{-1}.

References

1. American College of Sports Medicine. ACSM's guidelines for exercise testing and prescription. 5th ed. Baltimore: Williams & Wilkins, 1995:278–283.

2. Brouha L. The step test: a simple method of measuring physical fitness for muscular work in young men. Res Quart 1943;14:31–36.

3. Sharkey BJ. Physiology of fitness. Champaign, IL: Human Kinetics, 1984:258–260.

Sample Problems Using the Stepping Equation

1. The Forestry step test for men consists of stepping on a 15.75-inch bench at a rate of 22.5 steps per minute. What is the expected VO$_2$ in ml·min^{-1}·kg^{-1} while performing this test?

2. The Forestry step test for women consists of stepping on a 13-inch bench at a rate of 22.5 steps per minute. What is the expected VO$_2$ in ml·min^{-1}·kg^{-1} while performing this test?

3. The Harvard step test for men consists of stepping on a 20-inch bench at a rate of 30 steps per minute. What is expected VO$_2$ in ml·min^{-1}·kg^{-1} while performing this test?

4. You want your client to exercise at 32 ml·min^{-1}·kg^{-1} while stepping on an 18-inch bench. What step rate should you use?

5. You want your client to exercise at 15 ml·min^{-1}·kg^{-1}. You plan to use a stepping rate of 24 steps per minute. What step height in inches is needed?

6. Your 66-kg client is stepping on a 15-inch bench at 30 steps per minute. What is her energy expenditure in kcal·min^{-1}?

7. You have 6-inch benches in your step aerobics class. During slow cool-down music, the class steps at 10 steps per minute. What VO_2 does this elicit in ml·min^{-1}·kg^{-1}?

8. You teach a step aerobics class that uses 2-inch stepping platforms that can be added on top of each other. Your music has a beat set at 120 per minute, so you plan to have your class step at 30 steps per minute. Assuming that most people in your class have an average capacity of 10 METs, how many step platforms should they use to exercise at 70% of their capacity?

9. You supervise several aerobics dance instructors. One instructor demonstrates her new step aerobics routine that moves from a 4-inch bench to a 10-inch bench while stepping at 40 steps per minute. What is the increase in VO_2 in ml·min^{-1}·kg^{-1} when going from the low to the high bench?

10. You want a 78-kg client to exercise at 1.6 L·min^{-1} while stepping on a 13-inch bench. At what rate should the metronome be set?

Answers to the Stepping Equation Sample Problems

1. Answer: 29.5 ml·min^{-1}·kg^{-1}

   Solution:

   $VO_2 = 0.35(22.5) + 0.061(22.5)(15.75)$

   $VO_2 = 7.875 + 21.617$

   $VO_2 = 29.49$

2. Answer: 25.7 ml·min^{-1}·kg^{-1}

   Solution:

   $VO_2 = 0.35(22.5) + 0.061(22.5)(13)$

   $VO_2 = 7.875 + 17.8425$

   $VO_2 = 25.72$

3. Answer: 47.1 ml·min^{-1}·kg^{-1}

   Solution:

   $VO_2 = 0.35(30) + 0.061(30)(20)$

   $VO_2 = 10.5 + 36.6$

   $VO_2 = 47.1$

4. Answer: 22 steps·min^{-1}

   Solution:

   $32 = 0.35(\text{rate}) + 0.061(\text{rate})(18)$

   $32 = 0.35(\text{rate}) + 1.098(\text{rate})$

   $32 = (0.35 + 1.098)(\text{rate})$

   $32 = 1.448(\text{rate})$

   $32/1.448 = \text{rate}$

   $22.099 = \text{rate}$

5. Answer: 4.5 inches

   Solution:

   $15 = 0.35(24) + 0.061(24)(\text{height})$

   $15 = 8.4 + 1.464(\text{height})$

   $15 - 8.4 = 1.464(\text{height})$

   $6.6 = 1.464(\text{height})$

   $6.6/1.464 = \text{height}$

   $4.508 = \text{height}$

6. Answer: 12.5 kcal·min^{-1}

   Solution:

   $VO_2 = 0.35(30) + 0.061(30)(15)$

   $VO_2 = 10.5 + 27.45$

   $VO_2 = 37.95$

   Convert to L·min^{-1}: $37.95 \times 66/1000 = 2.5047$ L·min^{-1}

   Convert to kcal·min^{-1}: $2.5047 \times 5 = 12.52$ kcal·min^{-1}

7. Answer: 7.2 ml·min^{-1}·kg^{-1}

   Solution:

   $VO_2 = 0.35(10) + 0.061(10)(6)$

   $VO_2 = 3.5 + 3.66$

   $VO_2 = 7.16$

8. Answer: Four 2-inch platforms

   Solution: The VO_2 is 0.70×10 METs $\times 3.5 = 24.5$ ml·min^{-1}·kg^{-1}. Enter this into the equation and solve for the height.

   $24.5 = 0.35(30) + 0.061(30)(\text{height})$

   $24.5 = 10.5 + 1.83(\text{height})$

   $24.5 - 10.5 = 1.83(\text{height})$

   $14.0 = 1.83(\text{height})$

   $14.0/1.83 = \text{height}$

   $7.65 = \text{height}$

   The total height needed is almost 8 inches. Because each platform is 2 inches high, you need $8/2 = 4$ platforms.

9. Answer: 14.6 ml·min^{-1}·kg^{-1}

   Solution: There are two ways to solve this problem. We'll do it the long way first, by determining the VO_2 for both levels of exercise and then taking the difference.

   VO_2 at 10 inches $= 0.35(40) + 0.061(40)(10)$

   VO_2 at 10 inches $= 14.0 + 24.4$

   VO_2 at 10 inches $= 38.4$

   VO_2 at 4 inches $= 0.35(40) + 0.061(40)(4)$

   VO_2 at 4 inches $= 14.0 + 9.76$

   VO_2 at 4 inches $= 23.76$

The increase in VO_2 that the routine requires when going from the 4-inch to the 10-inch bench is $38.4 - 23.76 = 14.64$ ml·min⁻¹·kg⁻¹. Alternatively, we could have solved this problem by realizing that the increase in VO_2 is due to a 6-inch increase in step height. Because no change in stepping rate occurred, the only change is in the last term (the vertical component) of the equation. Thus: $0.061(40)(6) = 14.64$ ml·min⁻¹·kg⁻¹.

10. Answer: 72 beats·min⁻¹

Solution: Convert the absolute VO_2 to relative terms: $1.6 \times 1000/78 = 20.5$ ml·min⁻¹·kg⁻¹. Now solve for the stepping rate.

$20.5 = 0.35(rate) + 0.061(rate)(13)$

$20.5 = 0.35(rate) + 0.793(rate)$

$20.5 = (0.35 + 0.793)(rate)$

$20.5 = 1.143(rate)$

$20.5/1.143 = rate$

$17.9 = rate$

The stepping rate is 18 steps·min⁻¹. The metronome rate must be set at 4 times this value: $18 \times 4 = 72$ beats·min⁻¹.

CHAPTER 5

OTHER MODES OF EXERCISE

Many exercise facilities have machines for cardiopulmonary training that are not covered by the ACSM equations. However, most of these machines provide reasonably accurate indications of workload that can be used to estimate VO_2. In this chapter, we will examine how to do this for a few of the better known machines. In addition, we will explain how to estimate the VO_2 for outdoor bicycling.

Stair Steppers

A variety of machines use stair stepping, with the most popular mode being stepping up and down on a pair of independently moving pedals. This form of exercise is quite different from the stepping described in Chapter 4—there is no eccentric component when stepping on the machine. The exerciser lifts each leg by concentrically activating the hip flexors, and pushes down each leg by concentrically activating the hip and knee extensors (and to a lesser extent, the plantar flexors of the ankle). This is similar to climbing a real flight of stairs. However, real stair climbing involves forward motion, whereas only vertical work is done on stair stepping machines. It would seem appropriate to modify the stepping equation in Chapter 4 by eliminating the "horizontal" component and by reducing the conversion factor in the vertical component to eliminate the eccentric work. However, when this is done, the result appears to overestimate the actual VO_2.

In 1992, Howley et al. (3) studied the VO_2 of stepping on a Stairmaster 4000. They found that the true MET level was about 20% less than that indicated on the machine. Since then, the Stairmaster company has revised the reported values on the machine to approximate true MET levels. Thus, if your client is exercising on a newer Stairmaster (that indicates work rates up to level 14), the work rate setting can be used as METs. If an older Stairmaster is being used (that indicates work rates up to 17), then deduct 20% from the setting.

For the newer models, use the following equation to get VO_2 in the units $ml \cdot min^{-1} \cdot kg^{-1}$:

STAIRMASTER

$$VO_2 = 3.5 \text{ (machine setting)}$$
$$ml \cdot min^{-1} \cdot kg^{-1} \qquad METs$$

Problem 1

Your 82-kg client is exercising on a Stairmaster at setting 6. What is his energy expenditure in $kcal \cdot min^{-1}$?

$VO_2 = 3.5(\text{machine setting})$

$VO_2 = 3.5(6)$

$VO_2 = 21.0$

The answer is 21.0 $ml \cdot min^{-1} \cdot kg^{-1}$. To get $kcal \cdot min^{-1}$, we must convert to $L \cdot min^{-1}$ and then to $kcal \cdot min^{-1}$ ($21.0 \times 82/1000 = 1.722$ $L \cdot min^{-1}$; $1.722 \times 5 = 8.61$ $kcal \cdot min^{-1}$). The answer is approximately 8.6 $kcal \cdot min^{-1}$.

It is common to observe clients placing their hands on the handrails and then locking their elbows, supporting much of their weight with their arms and shoulders instead of their legs. This greatly reduces the effort needed to perform the stepping and therefore the VO_2. They should be taught to rest their hands lightly on the handrails for balance only.

Rowing Machines

When used properly, rowing machines are an excellent form of upper body plus lower body exercise. Because the arms are used, it is tempting to apply the arm ergometry equation to calculate the VO_2 from the workload displayed on the rowing machine. This would not be correct. The arms are less efficient than the legs because the arms have less muscle mass. The more muscle you use, up to a point, in performing any exercise, the more efficient you become. In rowing, a sufficient amount of muscle mass is activated to achieve optimal efficiency. Therefore, although the arms are being used, the **leg** ergometry equation should provide accurate estimates of VO_2.

The accuracy of this analysis was confirmed by Hagerman et al. (2), who measured the VO_2 of subjects exercising on a Concept II rower. Based on their data, approximately 2 ml·min^{-1} were needed for each kg·m·min^{-1} of workload, just as it is in the ACSM leg ergometry equation.

ROWING

$$VO_2 = 3.5 + 2(\text{workload})/\text{BW}$$
ml·min^{-1}·kg^{-1} \qquad kg·m·min^{-1} \quad kg

Problem 2
Your 43-kg client is rowing a Concept II rower at 150 watts. What is her VO_2 in ml·min^{-1}·kg^{-1}?

$VO_2 = 3.5 + 2(\text{workload})/\text{BW}$

Set up the bike equation and solve. To express the workload in kg·m·min^{-1}, multiply the watts by 6: $150 \times 6 = 900$ kg·m·min^{-1}.

$VO_2 = 3.5 + 2(900)/43$

$VO_2 = 3.5 + 1800/43$

$VO_2 = 3.5 + 41.86$

$VO_2 = 45.36$

The answer is approximately 45.4 ml·min^{-1}·kg^{-1}.

Arm Plus Leg Bikes

Arm plus leg bikes allow the client to exercise with the arms alone in a push/pull motion, with the legs alone in normal cycling, or with the arms and legs together. Based on the discussion of muscle mass and efficiency above, you can probably guess how to apply the ACSM equations to this type of machine. When the legs are being used alone or when the arms are being used with the legs, you exercise efficiently. Thus, the leg ergometry equation, with its 2 times workload conversion factor, should be used to estimate VO_2. When the arms are used by themselves, you are less efficient. Thus, the arm ergometry equation, with a conversion factor of 3 × workload, should be used instead.

This was confirmed by Hagan et al. (1), who measured the VO_2 of subjects exercising on a Schwinn Airdyne. Based on their data, approximately 2 ml·min^{-1} were needed for each kg·m·min^{-1} when the subjects exercised with legs alone or with arms plus legs. Approximately 3 ml·min^{-1} were needed for each kg·m·min^{-1} when they exercised with arms alone. The Schwinn Airdyne reports workload as level 1, level 2, etc. Each level is approximately equal to 300 kg·m·min^{-1}.

ARM BIKE

$$VO_2 = 3.5 + 3(\text{workload})/BW$$
ml·min^{-1}·kg^{-1} kg·m·min^{-1} kg

LEG BIKE

$$VO_2 = 3.5 + 2(\text{workload})/BW$$
ml·min^{-1}·kg^{-1} kg·m·min^{-1} kg

ARM + LEG BIKE

$$VO_2 = 3.5 + 2(\text{workload})/BW$$
ml·min^{-1}·kg^{-1} kg·m·min^{-1} kg

Problem 3

Your 73-kg client is exercising on a Schwinn Airdyne at level 2, switching back and forth between arms alone and arms plus legs. What is his expected VO_2 in ml·min^{-1}·kg^{-1} during these two modes of exercise?

$VO_2 = 3.5 + 3(\text{workload})/BW$

Set up the equation. Let's do the arms first. Fill in the unknowns and solve. Because there are 300 kg·m·min^{-1} for each level, the workload is 2 × 300 = 600 kg·m·min^{-1}.

$VO_2 = 3.5 + 3(600)/73$

$VO_2 = 3.5 + 1800/73$

$VO_2 = 3.5 + 24.66$

$VO_2 = 28.16$

$VO_2 = 3.5 + 2(\text{workload})/BW$

Now, let's reset the equation for the arms plus legs. The workload is the same.

$VO_2 = 3.5 + 2(600)/73$

$VO_2 = 3.5 + 1200/73$

$VO_2 = 3.5 + 16.44$

$VO_2 = 19.94$

The answer is approximately 28.2 ml·min^{-1}·kg^{-1} with the arms and approximately 19.9 ml·min^{-1}·kg^{-1} with arms plus legs.

Some models of the Schwinn Airdyne report the workload in watts and in METs (the latter requires you to enter your body weight). It is critical to recognize that the bike's computer does not know if you are exercising with arms, legs, or both. The MET read-out is based on leg (or arm plus leg) exercise. Thus, if you are exercising with arms alone, the MET readout will be incorrect; the true MET level would be 50% greater than what the readout claims.

Outdoor Bicycling

The energy cost of cycling outdoors is quite complex, being affected not only by the road speed of the cyclist, but also by the wind speed, hills, type and weight of the bicycle, position of the cyclist on the bike (upright versus tucked down), and whether the cyclist is "drafting" other riders (following close enough to reduce the air drag). There are equations in the scientific literature that allow one to compute the effects of these factors, but they are too complex for general exercise prescription purposes.

Fortunately, a fairly good estimate of the VO_2 during outdoor bicycling can be made for riding on flat ground in relatively still air. Based on a study by Swain et al. (4), Table 5.1 provides the net VO_2 in ml·min^{-1} for cycling at different speeds. As discussed in Chapter 3, the net VO_2 in absolute terms during stationary cycling on an ergometer is the same regardless of the client's size. For outdoor cycling, there is a small added cost for larger cyclists to push their bodies through the air. Thus, when using Table 5.1, choose the value for large or small cyclists based on your client's size. Large refers to cyclists who weigh approximately 85 kg; small refers to cyclists who weigh about 60 kg. Because there is only a small difference in the net VO_2, it is sufficient to place your client into the category to which he or she is closest.

In the study by Swain et al. (4), the cyclists were on lightweight road bikes and they placed their hands on the dropped portion of the handlebars. Bicycling on heavy mountain bikes or with a more upright posture would raise the VO_2. Riding with aerodynamic handlebars would lower the VO_2. Finally, the study measured the VO_2 while cycling from 10 to 20 mph. Thus, the listed VO_2s for speeds above and below that range are estimates.

All equations in this book for VO_2 during exercise yield relative VO_2 in the units ml·min^{-1}·kg^{-1}. To determine the VO_2 in those units for outdoor bicycling, use the following equation:

OUTDOOR CYCLING

$$VO_2 = 3.5 + \text{(tabled value)}/BW$$
ml·min^{-1}·kg^{-1} \qquad ml·min^{-1} \qquad kg

	VO$_2$ (ml·min^{-1})	
TABLE 5-1.		
Net VO$_2$ During Outdoor Bicycling		
Speed (mph)	Small Cyclists	Large Cyclists
5	450	460
6	510	520
7	580	590
8	660	670
9	750	770
10	850	880
11	960	1000
12	1080	1130
13	1220	1280
14	1360	1440
15	1510	1610
16	1680	1790
17	1860	1990
18	2040	2190
19	2240	2410
20	2450	2650
21	2670	2890
22	2900	3150
23	3140	3410
24	3390	3700
25	3660	3990
26	3930	4290
27	4210	4610
28	4510	4940
29	4820	5280
30	5130	5640

Problem 4

A 68-kg cyclist and a 90-kg cyclist are riding side by side at 20 mph. What are their estimated VO$_2$s in ml·min^{-1}·kg^{-1}?

VO$_2$ = 3.5 + (tabled value)/BW

Let's do the smaller cyclist first. From Table 5.1, the net VO$_2$ of a small cyclist at 20 mph is approximately 2450 ml·min^{-1}.

VO$_2$ = 3.5 + (2450)/68

VO$_2$ = 3.5 + 36.03

VO$_2$ = 39.53

VO$_2$ = 3.5 + (2650)/90

Now, let's set it up for the larger cyclist. According to Table 5.1, his net VO$_2$ is approximately 2650 ml·min^{-1}·kg^{-1}.

VO$_2$ = 3.5 + 29.44

VO$_2$ = 32.94

The answer is approximately 40 ml·min^{-1}·kg^{-1} for the 68-kg cyclist and approximately 33 ml·min^{-1}·kg^{-1} for the 90-kg cyclist. Note that outdoor riding is harder for smaller individuals, at least on flat ground.

In the fitness setting, it is too complicated to make adjustments to the bicycling VO$_2$ for hills, wind, or drafting. However, the client can use heart rate to provide equivalencies. Suppose you have established a target heart rate of 145 to 150 bpm for your client, and that the client achieves this heart rate while cycling on flat ground in still air at 17 mph. It is safe to assume that if the client maintains the same heart rate when going up hill or into a headwind, the VO$_2$ is unchanged although the road speed must drop. The only confounding factor is if the client becomes overheated. During any form of exercise, overheating will raise heart rate. Thus, it will no longer provide an accurate reflection of the VO$_2$.

References

1. Hagan RD, Gettman LR, Upton SJ, et al. Cardiorespiratory responses to arm, leg, and combined arm and leg work on an air-braked ergometer. J Cardiac Rehabil 1983;3:689–695.

2. Hagerman FC, Lawrence RA, Mansfield MC. A comparison of energy expenditure during rowing and cycling ergometry. Med Sci Sports Exerc 1988;20:479–488.

3. Howley ET, Colacino DL, Swensen TC. Factors affecting the oxygen cost of stepping on an electronic stepping ergometer. Med Sci Sports Exerc 1992;24:1055–1058.

4. Swain DP, Coast JR, Clifford PS, et al. Influence of body size on oxygen consumption during bicycling. J Appl Physiol 1987;62:668–672.

Sample Problems Using Other Modes of Exercise

1. Your client is exercising at level 10 on a Stairmaster. What is her VO$_2$ in ml·min^{-1}·kg^{-1}?

2. Your 67-kg client is exercising at level 7 on a Stairmaster. What is her energy expenditure in kcal·min^{-1}?

3. You want your client to exercise at 70% of her 46 ml·min^{-1}·kg^{-1} capacity. What setting on a Stairmaster would suffice?

4. Your 89-kg client is exercising at 230 watts on a Concept II rower. What is his VO$_2$ in ml·min^{-1}·kg^{-1}?

5. You want your 42-kg client to exercise at 7 METs on a Concept II rower. What intensity in watts should she use?

6. Your 58-kg client exercised for 30 minutes on a Concept II rower at an average power of 176 watts. How many kilocalories did she burn?

7. Your 63-kg client is exercising on a Schwinn Airdyne with arms and legs together at level 4. What is his VO$_2$ in L·min^{-1}?

8. Your 70-kg client has a peak VO$_2$ during arm exercise of 22 ml·min^{-1}·kg^{-1}. At what level on a Schwinn Airdyne should he exercise to achieve 60% of this?

9. Your 82-kg client is exercising on a Schwinn Airdyne with legs alone at an average level of 2.7. What is his VO$_2$ in ml·min^{-1}·kg^{-1}?

10. You want your 60-kg client to exercise at 7 METs. At what speed should she ride her bicycle outdoors?

Answers to Sample Problems Using Other Modes of Exercise

1. Answer: 35.0 ml·min^{-1}·kg^{-1}

   Solution:

   $VO_2 = 3.5(10)$

   $VO_2 = 35.0$

2. Answer: 8.2 kcal·min^{-1}

   Solution:

   $VO_2 = 3.5(7)$

   $VO_2 = 24.5$

   Multiply by body weight and divide by 1000 to get absolute VO_2: 24.5 × 67/1000 = 1.6415 L·min^{-1}. Multiply by 5 to get caloric expenditure: 1.6415 × 5 = 8.2075 kcal·min^{-1}.

3. Answer: setting 9

   Solution: The desired VO_2 is 0.70 × 46 = 32.2 ml·min^{-1}·kg^{-1}.

   32.2 = 3.5(machine setting)

   32.2/3.5 = 9.2 METs

4. Answer: 34.5 ml·min^{-1}·kg^{-1}

   Solution: Watts times 6 yields the workload in appropriate units: 230 × 6 = 1380 kg·m·min^{-1}. Enter into leg ergometry equation and solve.

   $VO_2 = 3.5 + 2(1380)/89$

   $VO_2 = 3.5 + 2760/89$

   $VO_2 = 3.5 + 31.01$

   $VO_2 = 34.51$

5. Answer: approximately 74 watts

   Solution: 7 METs × 3.5 = 24.5 ml·min^{-1}·kg^{-1}. Enter into leg ergometry equation and solve.

   24.5 = 3.5 + 2(workload)/42

   24.5 − 3.5 = 2(workload)/42

   21.0 = 2(workload)/42

   21.0 × 42/2 = workload

   441 = workload

   The answer is 441 kg·m·min^{-1}. Divide by 6 to get watts: 441/6 = 73.5 watts.

6. Answer: approximately 348 kcal

   Solution: To get to kilocalories, we need VO_2. Let's use the leg ergometry equation to get relative VO_2 and then convert it to absolute units. The workload is 176 watts × 6 = 1056 kg·m·min^{-1}.

   $VO_2 = 3.5 + 2(1056)/58$

   $VO_2 = 3.5 + 2112/58$

   $VO_2 = 3.5 + 36.41$

   $VO_2 = 39.91$

The VO_2 is about 39.9 ml·min^{-1}·kg^{-1}. Multiply by body weight and divide by 1000 to get absolute units: 39.9 × 58/1000 = 2.314 L·min^{-1}. Now, multiply by 5 to get kcal·min^{-1}: 2.314 × 5 = 11.6 kcal·min^{-1}. This is the rate per minute. The total burned in 30 minutes is: 11.6 × 30 = 348 kcal.

7. Answer: 2.62 L·min^{-1}

   Solution: Level 4 × 300 = 1200 kg·m·min^{-1}. Enter this into the leg ergometry equation and solve.

   VO_2 = 3.5 + 2(1200)/63

   VO_2 = 3.5 + 2400/63

   VO_2 = 3.5 + 38.10

   VO_2 = 41.60

   Multiply by body weight and divide by 1000 to get absolute VO_2: 41.6 × 63/1000 = 2.62 L·min^{-1}.

8. Answer: approximately level 0.8

   Solution: The desired VO_2 is 0.60 × 22 = 13.2 ml·min^{-1}·kg^{-1}. Enter into the **arm** ergometry equation and solve.

   13.2 = 3.5 + 3(workload)/70

   13.2 − 3.5 = 3(workload)/70

   9.7 = 3(workload)/70

   9.7 × 70/3 = workload

   226 = workload

   The answer is 226 kg·m·min^{-1}. Divide by 300 to get the Airdyne level: 226/300 = 0.75.

9. Answer: 23.3 ml·min^{-1}·kg^{-1}

   Solution: Level 2.7 × 300 = 810 kg·m·min^{-1}. Enter this into the leg ergometry equation and solve.

   VO_2 = 3.5 + 2(810)/82

   VO_2 = 3.5 + 1620/82

   VO_2 = 3.5 + 19.76

   VO_2 = 23.26

10. Answer: approximately 13 mph

   Solution: 7 METs × 3.5 = 24.5 ml·min^{-1}·kg^{-1}. Enter into the outdoor cycling equation and solve.

   24.5 = 3.5 + (tabled value)/60

   24.5 − 3.5 = (tabled value)/60

   21.0 = (tabled value)/60

   21.0 × 60 = tabled value

   1260 = tabled value

   Our small cyclist is looking for a net VO_2 of 1260 ml·min^{-1}. According to Table 5.1, a speed of 13 mph yields a net VO_2 of about 1220 ml·min^{-1}. Close enough.

CHAPTER 6

CALCULATING A HEART RATE PRESCRIPTION

% HEART RATE RESERVE METHOD

Target HR = (HRRfraction)(HRmax − HRrest) + HRrest

There are many facets to an exercise prescription. These facets are best summarized by the FITT principle:

Frequency: how often the client exercises; generally expressed in sessions per week or sometimes in sessions per day.

Intensity: how hard the client exercises; generally expressed as a percentage of maximum effort.

Time: the duration of each exercise session.

Type: the mode of exercise; for example, if the exercise is aerobic, whether it is to be performed by walking, running, cycling, or stepping.

For aerobic exercises, the intensity of effort is expressed as a percentage of maximal VO_2. If the client's VO_2max is known, the intensity can be calculated as a workload from the equations in Chapters 1–5. However, the VO_2max is often unknown. Under such circumstances, the intensity can be prescribed by heart rate. This is based on the well-established linear relationship between oxygen consumption and heart rate. This chapter explains the mathematical procedures involved in calculating a target heart rate as part of an overall exercise prescription. Other aspects of designing an appropriate exercise prescription are covered in other resources, such as the 5th edition of the "ACSM's Guidelines for Exercise Testing and Prescription" (1).

The ACSM recommends that intensity be set at 50 to 85% of VO_2max for improving cardiorespiratory fitness, with a narrower range selected for any one client. For example, a sedentary individual who is just beginning an exercise program might be placed at 50 to 60%. For severely deconditioned subjects and special clinical populations, an intensity as low as 40% may be used.

There are three methods for determining a target heart rate prescription after an intensity range has been selected: 1) the maximal graded exercise test method, 2) the percentage of heart rate reserve method, and 3) the percentage of maximal heart rate method.

The first method can only be used if the patient has undergone a maximal graded exercise test with the use of a metabolic cart to directly measure VO_2. After the maximal test is completed, one reviews the data to find the heart rate that occurred at the desired level of VO_2. Because this method is beyond the scope of most practitioners, it will not be covered further. The remainder of this chapter will explain the use of the other two methods of determining a target heart rate, the percentage of heart rate reserve method and the percentage of maximum heart rate method.

Percentage of Heart Rate Reserve Method

The heart rate reserve method of calculating an exercise heart rate was first described by Karvonen et al. (2). Thus, it is often referred to as the Karvonen method. A close relationship exists between oxygen consumption and the *difference* between maximum and resting heart rate. This difference (HRmax − HRrest) is known as heart rate reserve (HRR). The HRR method assumes that when a person is exercising at 50% of VO$_2$max, he or she is at 50% of HRR; when at 60% of VO$_2$max, the person is at 60% of HRR; and so on. This assumption is not correct, but it is fairly accurate for individuals with functional capacities above 10 METs. For low fit clients, this assumption introduces a sizable error, as demonstrated in recent studies (3, 4). For complete accuracy, exercise prescriptions should be based on net VO$_2$ instead of gross VO$_2$, because a person exercising at 50% of HRR is at 50% of VO$_2$ reserve, i.e., halfway between resting and maximum VO$_2$, not at 50% of maximum VO$_2$. In this text, we will follow the ACSM prescription practice of referring to the intensity as a percentage of VO$_2$max.

To determine a target heart rate, you must know the client's maximum heart rate and resting heart rate. Maximum heart rate is defined as the highest heart rate attained during a maximal graded exercise test. When this information is not available, it is common to estimate the maximum heart rate as 220 minus the client's age in years. Thus, a 45-year-old client would be expected to have an HRmax of $220 - 45 = 175$ bpm. There is considerable variation in this value. Approximately one third of the population can be expected to have a true HRmax that is more than 10 bpm above or below this estimate. Thus, it is much better to determine the true HRmax than to estimate it. If HRmax is estimated, the target heart rate that is calculated from it should be considered a starting point. The practitioner should carefully observe the client as the target heart rate is approached and be prepared to increase or decrease the target value if the exercise appears to be too easy or too difficult.

The resting heart rate should be determined when the client has been in a state of quiet rest for several minutes. The client should not be overheated, dehydrated, or under the influence of recently consumed caffeine, nicotine, or other drugs that influence heart rate.

Problem 1

Your 37-year-old client has a maximum heart rate of 196 bpm and a resting heart rate of 64 bpm. Using the HRR method, what would be her target heart rate range at 60 to 70% of functional capacity?

target HR = (HRR fraction)(HRmax − HRrest) + HRrest	Set up the target heart rate equation and enter the known values. Intensity must be entered as a fraction, i.e., 0.60 for 60% and 0.70 for 70%. We will use 0.60 first and then solve the equation a second time using 0.70. Use the known HRmax rather than estimating it.
target HR = (0.60)(196 − 64) + 64	In solving this equation, first perform functions that are inside parentheses, second perform multiplication of terms, and third perform addition of terms.
target HR = (0.60)(132) + 64	The HRR is 132 bpm. This is how much the heart rate can increase above rest. We want it to increase 60% above rest.
target HR = 79.2 + 64	We want the heart rate to increase 79 bpm above rest. Now we add the resting value back.

target HR = 143.2

target HR = (0.70)(132) + 64   Reset the equation for the upper limit heart rate at 70%. Note that we already know the HRR value, 132, so we do not need to calculate it a second time from HRmax − HRrest.

target HR = 92.4 + 64

target HR = 156.4

The answer is 143 to 156 bpm. This heart rate range should place the client at close to 60 to 70% of VO_2max. (In fact, as explained earlier, the client would be at 60 to 70% of net VO_2, i.e., VO_2 reserve, not gross VO_2).

Problem 2

You want your 36-year-old client to exercise at 50% of his VO_2max. His resting heart rate is 72 bpm. Using the HRR method, what is his target heart rate?

target HR = (HRR fraction)(HRmax − HRrest) + HRrest   Set up the equation and enter the known values. HRmax is not known, so it must be estimated as 220 − age: 220 − 36 = 184 bpm.

target HR = (0.50)(184 − 72) + 72

target HR = (0.50)(112) + 72

target HR = 56 + 72

target HR = 128

The answer is 128 bpm.

Percentage of Maximum Heart Rate Method

A simpler means of calculating target heart rates is using a straight percentage of the maximal heart rate, without taking the resting heart rate into consideration. The HRR method is usually preferred, but it would be better to use a percentage of maximum heart rate if you do not have a reliable measurement of the client's resting heart rate.

It is critical to recognize that percentage of HRmax is not equivalent to percentage of VO_2max. At rest, a person is typically at about 10% of VO_2max (1 MET of a maximum of 10) but at 30 to 40% of HRmax (e.g., 70 bpm of a maximum of 200). The percentage of HRmax value is much higher than the percentage of VO_2max value at rest and gradually approaches the VO_2max value with increasing exercise intensity, because they are both 100% at maximum effort. Thus, if you want your client to exercise at 50% of VO_2max, it would be incorrect to place him or her at 50% of HRmax. According to the ACSM, it takes 60% of HRmax for most individuals to be at 50% of VO_2max, and 90% of HRmax to be at 85% of VO_2max.

Unfortunately, the ACSM values are not well supported in the scientific literature. A better equivalency between percentage VO_2max and percentage HRmax, based on a study by Swain et al. (5), is presented in Table 6.1.

TABLE 6-1.

% HRmax Values at Given Percentages of VO$_2$max

% VO$_2$max	% HRmax
40	63
50	70
60	76
70	82
80	89
85	92

To calculate a target heart rate using the percentage of maximum heart rate method, use the following formula:

% MAXIMUM HEART RATE METHOD

Target HR = (HRmax fraction)(HRmax)

Where the intensity fraction is from the desired percentage of maximum heart rate, not the desired percentage of VO$_2$max.

Problem 3

As in problem 2, you want a 36-year-old client to exercise at 50% of VO$_2$max. What is his target heart rate, using the percentage of maximum heart rate method?

target HR = (HRmax fraction)(HRmax)

Set up the equation and enter the known values. HRmax is estimated as 220 − 36 = 184 bpm. The intensity fraction is not 0.50, from the percentage VO$_2$max prescription, but must be raised to reflect the appropriate fraction of maximum heart rate. Let's use the ACSM value of 60%, or 0.60.

target HR = (0.60)(184)

target HR = 110.4

TABLE 6-2.

Heart Rate Conversion for 10 Seconds

Beats in 10 sec	Heart Rate (bpm)	Beats in 10 sec	Heart Rate (bpm)
8	48	22	132
9	54	23	138
10	60	24	144
11	66	25	150
12	72	26	156
13	78	27	162
14	84	28	168
15	90	29	174
16	96	30	180
17	102	31	186
18	108	32	192
19	114	33	198
20	120	34	204
21	126	35	210

TABLE 6-3.

Heart Rate Conversion for 15 Seconds

Beats in 15 sec	Heart Rate (bpm)	Beats in 15 sec	Heart Rate (bpm)
12	48	32	128
13	52	33	132
14	56	34	136
15	60	35	140
16	64	36	144
17	68	37	148
18	72	38	152
19	76	39	156
20	80	40	160
21	84	41	164
22	88	42	168
23	92	43	172
24	96	44	176
25	100	45	180
26	104	46	184
27	108	47	188
28	112	48	192
29	116	49	196
30	120	50	200
31	124	51	204

The ACSM answer is 110 bpm. If we use Table 6.1 to determine the percentage of HRmax, the answer would be $0.70 \times 184 = 128.8$, or approximately 129 bpm.

How will your clients know if they are at their target heart rates? Counting heart beats during exercise is generally not done for a full minute. It is simpler to count for 10 seconds and multiply the count by 6, or to count for 15 seconds and multiply the count by 4. Alternatively, the conversion charts in Tables 6.2 and 6.3 can be used.

References

1. American College of Sports Medicine. ACSM's guidelines for exercise testing and prescription. 5th ed. Baltimore: Williams & Wilkins, 1995:158–160.

2. Karvonen MJ, Kentala E, Mustala O. The effects of training on heart rate: a longitudinal study. Ann Med Exp Biol Fenn 1957;35:307–315.

3. Leutholtz BC, Swain DP, King ME, et al. Comparison of % heart rate reserve with %VO$_2$max and %VO$_2$ reserve during treadmill exercise (abstract). Med Sci Sports Exerc, in press.

4. Swain DP, Leutholtz BC. % Heart rate reserve is equivalent to %VO$_2$ reserve, not to %VO$_2$max. Med Sci Sports Exerc 1997;29:410–414.

5. Swain DP, Abernathy KS, Smith CS, et al. Target heart rates for the development of cardiorespiratory fitness. Med Sci Sports Exerc 1994;26:112–116.

Sample Problems in Calculating a Heart Rate Prescription

1. You want a 45-year-old client to exercise at 70% of her functional capacity. Her resting heart rate is 78 bpm. What is her target heart rate using the percentage of HRR method?

2. Your 76-year-old client has a maximum heart rate of 132 bpm and a resting heart rate of 72 bpm. What would be her target heart rate at 40% of VO_2max using the percentage of HRR method?

3. Your 25-year-old client has a resting heart rate of 64 bpm. What would be his target heart rate at 80% of VO_2max using the percentage of HRR method?

4. Your 37-year-old client has resting heart rate of 56 bpm. What would be her target heart rate at 75% of functional capacity using the percentage of HRR method?

5. Your 40-year-old client has a maximum heart rate of 198 bpm and a resting heart rate of 72 bpm. What would be his target heart rate at 60% of VO_2max using the percentage of HRR method?

6. You want your 30-year-old client to exercise at 70% of VO_2max. Using Table 6.1 and the percentage HRmax method, what would be his target HR?

7. You want your 62-year-old client to exercise at 60% of VO_2max. Using Table 6.1 and the percentage HRmax method, what would be his target HR?

8. You want your 44-year-old client to exercise at 65% of VO_2max. Using Table 6.1 and the percentage HRmax method, what would be his target HR?

9. Your 55-year-old client has a maximum heart rate of 156 bpm. Using Table 6.1 and the percentage HRmax method, what would be his target heart rate at 50% of VO_2max?

10. During an aerobic dance class, how many beats in 10 seconds should the average client obtain if they are trying to maintain a workout at 70% of VO_2max? Assume that the clients have an average age of 25 years. Use Table 6.1 and the percentage HRmax method.

Answers to Sample Problems in Calculating a Heart Rate Prescription

1. Answer: 146 bpm

    Solution:

    target HR = (0.70)[(220 − 45) − 78] + 78

    target HR = (0.70)(175 − 78) + 78

    target HR = (0.70)(97) + 78

    target HR = 67.9 + 78

    target HR = 145.9

2. Answer: 96 bpm

    Solution:

    target HR = (0.40)(132 − 72) + 72

    target HR = (0.40)(60) + 72

    target HR = 24 + 72

    target HR = 96

3. Answer: 169 bpm

    Solution:

    target HR = (0.80)[(220 − 25) − 64] + 64

    target HR = (0.80)(195 − 64) + 64

target HR = (0.80)(131) + 64

target HR = 104.8 + 64

target HR = 168.8

4. Answer: 151 bpm

Solution:

target HR = (0.75)[(220 − 37) − 56] + 56

target HR = (0.75)(183 − 56) + 56

target HR = (0.75)(127) + 56

target HR = 95.25 + 56

target HR = 151.25

5. Answer: 148 bpm

Solution:

target HR = (0.60)(198 − 72) + 72

target HR = (0.60)(126) + 72

target HR = 75.6 + 72

target HR = 147.6

6. Answer: 156 bpm

Solution: According to Table 6.1, 70% of VO_2max elicits 82% of HRmax.

target HR = (0.82)(220 − 30)

target HR = (0.82)(190)

target HR = 155.8

7. Answer: 120 bpm

Solution: According to Table 6.1, 60% of VO_2max elicits 76% of HRmax.

target HR = (0.76)(220 − 62)

target HR = (0.76)(158)

target HR = 120.08

8. Answer: 139 bpm

Solution: According to Table 6.1, 60% and 70% of VO_2max elicit 76% and 82% of HRmax, respectively. Interpolating between the two, 65% of VO_2max should elicit approximately 79% of HRmax.

target HR = (0.79)(220 − 44)

target HR = (0.79)(176)

target HR = 139.04

9. Answer: 109 bpm

Solution: According to Table 6.1, 50% of VO_2max elicits 70% of HRmax.

target HR = (0.70)(156)

target HR = 109.2

10. Answer: 27 beats

Solution: According to Table 6.1, 70% of VO_2max elicits 82% of HRmax.

target HR = (0.82)(220 − 25)

target HR = (0.82)(195)

target HR = 159.9

The target heart rate is 160 bpm. The number of beats that occurs in 10 seconds is one sixth this value: 160/6 = 26.7. Alternatively, look at Table 6.2 and find the 10-second count that most closely corresponds to the target heart rate.

CHAPTER 7

CALCULATING VO₂MAX

Calculating VO$_2$max from Maximal Exercise Tests

Knowing a client's VO$_2$max is very useful in designing an exercise prescription and in tracking the client's progress. For accuracy, it is best to measure the VO$_2$max during a maximal exercise test. However, according to ACSM guidelines, a physician must be present for maximal testing on all clients except those who are young (no more than 40 years of age for men, 50 years of age for women) and apparently healthy (no known cardiopulmonary or metabolic disease, no signs or symptoms of such disease, and no more than one major coronary risk factor).

In laboratory settings, the client performs a graded exercise test on a treadmill or cycle ergometer to the point of complete exhaustion. A metabolic cart is used to measure the client's ventilation, expired O_2 and CO_2 concentrations, and to calculate the VO$_2$max from these data. Metabolic carts are expensive and must be operated by trained personnel. In settings without a metabolic cart, one can still obtain an accurate determination of VO$_2$max from the final workload achieved during maximal testing.

Treadmill Maximal Testing

Foster et al. (3) published a regression equation that allows one to convert time (in minutes) on the Bruce protocol treadmill test to VO$_2$max with a correlation coefficient of 0.98 and a standard error of estimate (SEE) of ± 3.4 ml·min^{-1}·kg^{-1} (meaning that 68% of subjects had a true VO$_2$max within 3.4 units of the estimated value, and 95% were within 6.8 units). The equation is as follows:

$$VO_2max = 14.8 - 1.379(time) + 0.451(time^2) - 0.012(time^3)$$

Rather than using the equation, it is simpler to look up the VO$_2$max on the conversion chart that we have developed in Table 7.1. To ensure accurate results, the treadmill must be calibrated, and the client must avoid using the handrails during the test.

Cycle Ergometer Maximal Testing

Storer et al. (6) published a protocol for determining VO$_2$max from cycle ergometry. After a 4-minute warm up at 0 watts, the power is increased by 15 watts per minute. This can be done on a Monark bike by using 0.25-kg increments while pedaling at 60 rpm. The final completed power increment is then entered into one of the following gender-specific equations:

$$\text{males } VO_2max = [10.51(power) + 6.35(BW) - 10.49(age) + 519.3]/(BW)$$
$$\text{females } VO_2max = [9.39(power) + 7.7(BW) - 5.88(age) + 136.7]/(BW)$$

where: VO$_2$max is in ml·min^{-1}·kg^{-1}; power is in watts; BW is body weight in kg; and age is in years.

The equation for men has a correlation coefficient of 0.94, whereas that for women is 0.93. The SEEs were originally reported in L·min^{-1}; transposing to relative units, the SEEs were 2.6 and 2.3 ml·min^{-1}·kg^{-1} for men and women, respectively. To ensure accuracy, the ergometer must be calibrated, and the client must maintain the proper cadence. Because several variables are entered into the equations, it is not possible to derive a simple chart that covers all possible clients. Let's illustrate the use of the equations with a problem.

TABLE 7-1.

VO$_2$max from Bruce Protocol Times[a]

Time	VO$_2$max	Time	VO$_2$max	Time	VO$_2$max
5:00–5:14	18	10:00–10:14	35	15:00–15:14	56
5:15–5:29	19	10:15–10:29	36	15:15–15:29	57
5:30–5:44	19	10:30–10:44	37	15:30–15:44	58
5:45–5:59	20	10:45–10:59	38	15:45–15:59	59
6:00–6:14	21	11:00–11:14	39	16:00–16:14	60
6:15–6:29	21	11:15–11:29	40	16:15–16:29	60
6:30–6:44	22	11:30–11:44	41	16:30–16:44	61
6:45–6:59	23	11:45–11:59	42	16:45–16:59	62
7:00–7:14	24	12:00–12:14	43	17:00–17:14	63
7:15–7:29	24	12:15–12:29	44	17:15–17:29	64
7:30–7:44	25	12:30–12:44	45	17:30–17:44	65
7:45–7:59	26	12:45–12:59	46	17:45–17:59	66
8:00–8:14	27	13:00–13:14	47	18:00–18:14	67
8:15–8:29	28	13:15–13:29	48	18:15–18:29	67
8:30–8:44	29	13:30–13:44	49	18:30–18:44	68
8:45–8:59	30	13:45–13:59	50	18:45–18:59	69
9:00–9:14	31	14:00–14:14	51	19:00–19:14	69
9:15–9:29	32	14:15–14:29	53	19:15–19:29	70
9:30–9:44	33	14:30–14:44	54	19:30–19:44	71
9:45–9:59	34	14:45–14:59	55	19:45–19:59	71

[a]Where time is in min:sec, and VO$_2$max is in ml·min^{-1}·kg^{-1}.

Problem 1

A male client completes the 240-watts stage of the Storer cycling protocol. He is 48 years old and weighs 75 kg. What is his VO$_2$max in ml·min^{-1}·kg^{-1}?

VO$_2$max = [10.51(power) + 6.35(BW) − 10.49(age) + 519.3]/(BW)	Set up the equation for men and enter the known values.
VO$_2$max = [10.51(240) + 6.35(75) − 10.49(48) + 519.3]/(75)	First do the multiplications that are within the outside brackets.
VO$_2$max = [2522.4 + 476.25 − 503.52 + 519.3]/(75)	Now do the additions and subtractions within the outside brackets.
VO$_2$max = [3014.43]/(75)	The VO$_2$max in absolute terms is a little over 3 L·min^{-1}. Now we divide by body weight to get relative VO$_2$max.

VO$_2$max = 40.1924

The answer is approximately 40.2 ml·min^{-1}·kg^{-1}.

Calculating VO$_2$max from Submaximal Exercise Tests

It is often inconvenient, impractical, or potentially unsafe to perform maximal tests. Under such circumstances, submaximal exercise tests may be used to estimate a client's VO_2max. The heart rate obtained from a submaximal bout of exercise is used to predict the maximal workload and/or maximal VO_2 that the client would have achieved if he or she had been allowed to continue the test to maximal heart rate. This prediction relies on the close relationship between heart rate and oxygen consumption to predict VO_2max. The accuracy of these tests, however, is generally not as great as with maximal tests. This is mostly because the subject's maximal heart rate is not known.

There are many submaximal testing protocols used for estimating VO_2max, including walking tests, running tests, bench stepping, and cycle ergometry. Two very useful categories of tests will be covered here: the Rockport Walking Test and cycle ergometry tests.

Rockport Walking Test

Kline et al. (5) published a method of deriving VO_2max from a 1-mile walk. The study was supported by the Rockport Company (manufacturers of walking shoes); thus, the protocol has come to be known as the Rockport Walking Test. One or more clients walk 1 mile as briskly as possible. Each client's time to finish the mile is recorded, and heart rate is measured during the 15 seconds immediately after the walk is concluded, as validated in a later study by Wilkie et al. (8). The clients may not run and should be instructed to maintain an even pace. Speeding up as the end of the mile is approached would raise the heart rate and throw off the result. A potential source of major error is the measurement of heart rate. In group settings, the clients are often taught to measure it themselves, generally with poor results. The heart rate should be measured by experienced fitness testers or by a chest strap monitor. VO_2max is calculated from the following equation, which has a correlation coefficient of 0.85 and an SEE of $\pm$ 5.5 ml·min^{-1}·kg^{-1}:

$$VO_2max = [6965.2 + 20.02(BW) - 25.7(age) + 595.5(sex) -$$
$$224(time) - 11.5(HR)]/(BW)$$

where: VO_2max is in ml·min^{-1}·kg^{-1}; BW is body weight in kg; age is in years; sex is 1 for males, 0 for females; time is in minutes; and HR is heart rate in bpm.

Problem 2

Your 56-year-old, 82-kg, male client completes a 1-mile walk in 18 minutes and 4 seconds, with a heart rate count of 34 beats in 15 seconds. What is his estimated VO_2max in ml·min^{-1}·kg^{-1}?

$VO_2max = [6965.2 + 20.02(BW) - 25.7(age) + 595.5(sex) - 224(time) - 11.5(HR)]/(BW)$

First set up the equation and enter the known values. Time must be expressed in minutes, i.e., 18 + 4/60 = 18 + 0.07 = 18.07 minutes. Heart rate must be expressed in bpm, i.e., 34 in 15 sec × 4 = 136 in 60 sec = 136 bpm. Because the client is male, a multiplier of "1" is entered into the "sex" term.

$VO_2max = [6965.2 + 20.02(82) - 25.7(56) + 595.5(1) - 224(18.07) - 11.5(136)]/(82)$

Now do the multiplications within the large brackets.

$VO_2max = [6965.2 + 1641.64 - 1439.2 + 595.5 - 4047.68 - 1564]/(82)$

Now do the additions and subtractions within the brackets.

VO$_2$max = [2151.46]/(82)

The VO$_2$max in absolute terms is a little over 2.1 L·min^{-1}. We now divide by body weight to get the relative VO$_2$max.

VO$_2$max = 26.237

The answer is approximately 26.2 ml·min^{-1}·kg^{-1}.

Problem 3

Your 34-year-old, 58-kg, female client completes a 1-mile walk in 14 minutes and 43 seconds, with a heart rate count of 28 beats in 15 seconds. What is her estimated VO$_2$max in ml·min^{-1}·kg^{-1}?

VO$_2$max = [6965.2 + 20.02(BW) − 25.7(age) + 595.5(sex) − 224(time) − 11.5(HR)]/(BW)

First set up the equation and enter the known values. Time in minutes is 14 + 43/60 = 14.72. Heart rate is 28 × 4 = 112 bpm. Because the client is female, a multiplier of 0 is entered into the sex term, which will cause that term to drop out.

VO$_2$max = [6965.2 + 20.02(58) − 25.7(34) + 595.5(0) − 224(14.72) − 11.5(112)]/(58)

VO$_2$max = [6965.2 + 1161.16 − 873.8 + 0 − 3297.28 − 1288]/(58)

VO$_2$max = [2667.28]/(58)

VO$_2$max = 45.99

The answer is approximately 46.0 ml·min^{-1}·kg^{-1}.

Cycle Ergometry Testing—The Astrand Bike Test

Cycle ergometers have been in use for several decades as a convenient means of estimating aerobic fitness. The first and one of the most popular protocols was developed by Astrand and Rhyming (2) in 1954. The subject cycles for 6 minutes at a constant workload and with a cadence of 50 rpm. Heart rate is measured at the end of the fifth and sixth minutes. The workload in kg·m·min^{-1} and the average of the two heart rates are looked up on a nomogram (Fig. 7.1) to obtain a raw value for VO$_2$max in L·min^{-1}. The raw value must be multiplied by an age correction factor and then converted into ml·min^{-1}·kg^{-1}. The choice of workload is based on the apparent fitness of the client and the judgment of the tester. Statistical information regarding the accuracy of this method was not reported in 1954 in the same format as it is reported today.

The originally published version of the nomogram has several superfluous lines regarding bench stepping and other factors that may confuse the user. In the authors' experiences, many individuals using the nomogram locate the workload on the wrong line. To avoid this problem, we have removed the superfluous material in Figure 7.1 so that the nomogram can be used more easily.

Problem 4

A 34-year-old, 74-kg woman performs an Astrand bike test at 750 kg·m·min^{-1}. Her heart rates at the ends of the fifth and sixth minutes were 146 and 150 bpm. What is her VO$_2$max in ml·min^{-1}·kg^{-1}?

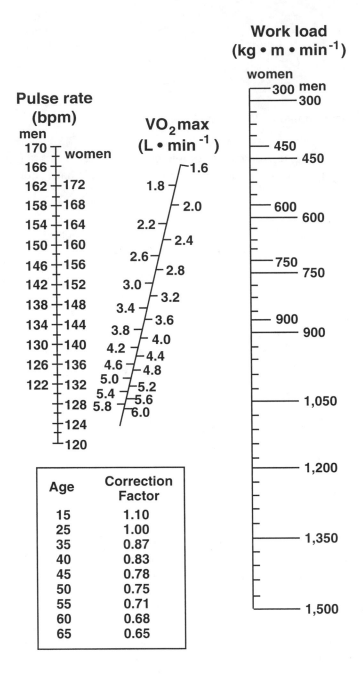

FIGURE 7.1.
Modified Astrand-Rhyming nomogram. With permission from Astrand PO, Rhyming IR. A nomogram for calculation of aerobic capacity (physical fitness) from pulse rate during submaximal work. J Appl Physiol 1954;7:218–221.

Her average heart rate for the test is 148 bpm. A straight-edge placed between 148 bpm on the women's heart rate scale and 750 kg·m·min^{-1} on the women's workload scale crosses the VO₂max scale at about 3.1 L·min^{-1}. The age correction factor for a 34 year old is approximately 0.88. Thus, her VO₂max is 3.1 × 0.88 = 2.728 L·min^{-1}. Converting to relative VO₂max: 2.728 × 1000/74 = 36.86 ml·min^{-1}·kg^{-1}.

The answer is approximately 37 ml·min^{-1}·kg^{-1}.

Cycle Ergometry Testing—The ACSM/YMCA Bike Test

Currently, the most commonly used bike test for estimating aerobic fitness follows the testing procedures described in the 5th edition of the "ACSM's Guidelines for Exercise Testing and Prescription" (1), which include the use of a workload protocol developed by the YMCA. The reader is referred to the ACSM guidelines for a detailed description of how to perform this test. Briefly, the client performs at least two stages of exercise at increasing workloads, for at least 3 minutes in each stage. The test is usually performed at a cadence of 50 rpm, but this is not necessary. Swain and Wright (7) recently demonstrated that tests performed at 80 rpm are equally valid. Thus, clients should be allowed to pick a cadence that is comfortable for them, with the resistance adjusted accordingly to provide appropriate workloads. The heart rates measured at the end of the third minute of each stage are used to predict VO₂max (provided they are within 6 bpm of the heart rates measured at the end of the second minute).

The accuracy of this type of test is not as great as the previous tests described in this chapter. Two published studies of this method obtained similar results. Greiwe et al. (4) obtained a correlation coefficient of 0.79 and SEE of $\pm$ 6.4 ml·min⁻¹·kg⁻¹; Swain and Wright (7) obtained a correlation coefficient of 0.81 and SEE of $\pm$ 7.4 ml·min⁻¹·kg⁻¹. Disturbingly, both studies found that the method overpredicted the true VO₂max by approximately 27%.

With any submaximal test used to estimate VO₂max, the primary purpose is to track the client's progress. If the test is performed in the same manner a second time, it will show if the client has changed his or her fitness level. If an accurate measurement of VO₂max is needed (e.g., as the basis of an exercise prescription), it is better to perform a maximal test.

The following procedures are used to convert the bike test data into a VO₂max value. First, heart rates from the end of each stage are plotted on a graph against their respective workload values. Heart rate is measured along the vertical (or *y*) axis, with workload on the horizontal (or *x*) axis. Then, a line of best fit is drawn through the data points. If several stages were completed by the client, it is appropriate to ignore the lower stages in drawing the line if the data from the lower stages do not fall along the same line as the upper stages. The line of best fit is extended beyond the data points so that it intersects a horizontal line at the level of the client's maximal heart rate. Of course, the maximal heart rate is probably not known, so it is estimated as 220 − age. Then, a vertical line is extended down from the intersection point to the *x*, or workload, axis. The estimated maximal workload is read from this point and converted to VO₂max using the leg cycle ergometry equation in Chapter 3. A generic graph (Fig. 7.2) that already has labeled *x* and *y* axes can be used for most clients. The projected maximal workloads of extremely fit clients will not fit on the generic graph. In such cases, simply construct a graph on plain graph paper using appropriate ranges for the axes.

Individuals using a calculator with statistical functions do not need to construct any graphs. Simply enter the workloads and heart rates as (*x,y*) pairs, then enter the estimated maximal heart rate as a new *y*, and press the function key for obtaining the corresponding *x'*. The resulting *x'* is the estimated maximal workload.

Problem 5

A 23-year-old, 92-kg client performs a submaximal bike test. His heart rates at the end of each stage are: 96 bpm at 150 kg·m·min⁻¹; 108 bpm at 450 kg·m·min⁻¹;

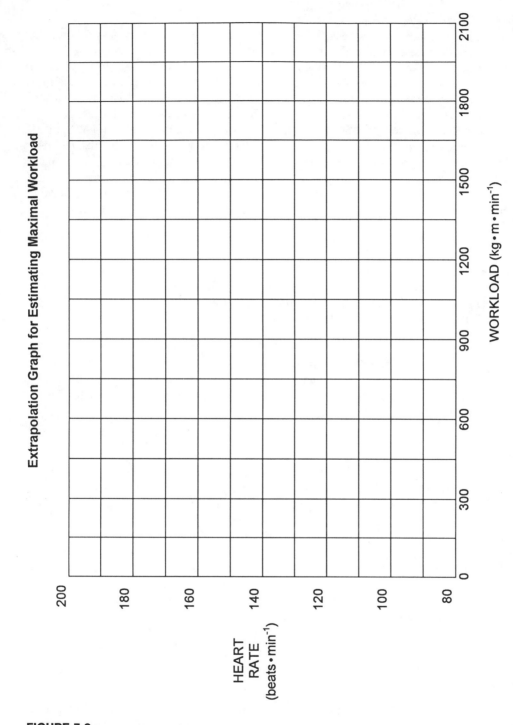

FIGURE 7.2.
Extrapolation graph for estimating maximal workload from heart rates and workloads recorded during a submaximal test.

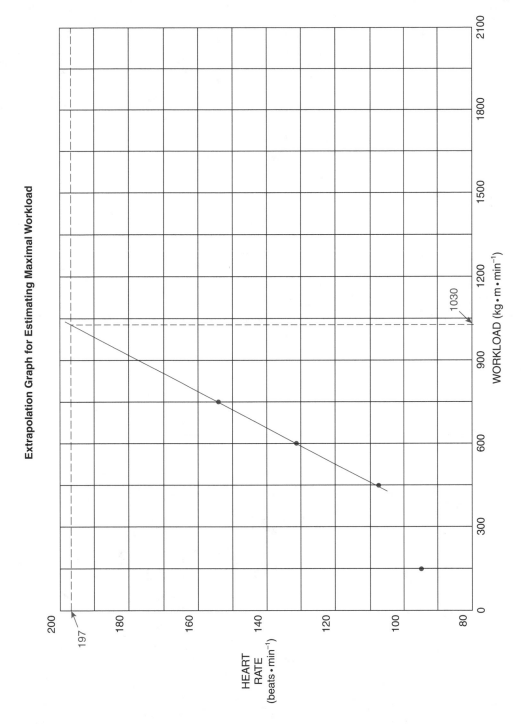

FIGURE 7.3.
Example of the estimation of a maximal workload, from problem 5.

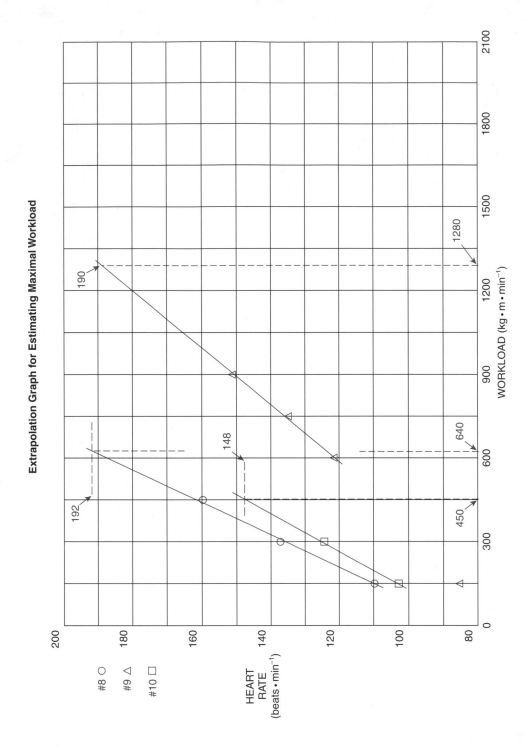

FIGURE 7.4.
Estimation of maximal workloads for sample problems 8, 9, and 10.

132 bpm at 600 kg·m·min^{-1}; and 154 bpm at 750 kg·m·min^{-1}. What is his VO$_2$max in ml·min^{-1}·kg^{-1}?

Plot the data as in Figure 7.3. Note that the last three data points fall along a fairly straight line. Therefore, the first data point will be ignored in drawing the line of best fit. When the line is extended upward to a heart rate of 197 bpm, it crosses at a workload of about 1030 kg·m·min^{-1}. This is the estimated maximum workload, i.e., the workload the client might have achieved if he had been allowed to continue the test to maximum. We now set up the leg ergometry equation and solve for VO$_2$max:

VO$_2$ = 3.5 + 2(workload)/BW
VO$_2$ = 3.5 + 2(1030)/92
VO$_2$ = 3.5 + 2060/92
VO$_2$ = 3.5 + 22.39
VO$_2$ = 25.89

The answer is approximately 26 ml·min^{-1}·kg^{-1}.

References

1. American College of Sports Medicine. ACSM's guidelines for exercise testing and prescription. 5th ed. Baltimore: Williams & Wilkins, 1995:12–19, 63–78.

2. Astrand PO, Rhyming IR. A nomogram for calculation of aerobic capacity (physical fitness) from pulse rate during submaximal work. J Appl Physiol 1954;7:218–221.

3. Foster C, Jackson AS, Pollock ML, et al. Generalized equations for predicting functional capacity from treadmill performance. Am Heart J 1984;107:1229–1234.

4. Greiwe JS, Kaminsky LA, Whaley MH, et al. Evaluation of the ACSM submaximal ergometer test for estimating VO$_2$max. Med Sci Sports Exerc 1995;27:1315–1320.

5. Kline GM, Porcari JP, Hintermeister R, et al. Estimation of VO$_2$max from a one-mile track walk, gender, age, and body weight. Med Sci Sports Exerc 1987;19:253–259.

6. Storer TW, Davis JA, Caiozzo VJ. Accurate prediction of VO$_2$max in cycle ergometry. Med Sci Sports Exerc 1990;22:704–712.

7. Swain DP, Wright RL. Prediction of VO$_2$peak from submaximal cycle ergometry using 50 versus 80 rpm. Med Sci Sports Exerc 1997;29:268–272.

8. Wilkie S, O'Hanley S, Ward A, et al. Estimation of VO$_2$max from a 1-mile walk test using recovery heart rate (abstract). Med Sci Sports Exerc 1987;19:168.

Sample Problems in Calculating VO$_2$max

1. Your 51-kg client completes 7 minutes and 47 seconds on the Bruce protocol treadmill test. What is her VO$_2$max in ml·min^{-1}·kg^{-1}, L·min^{-1}, and METs?

2. Your 32-year-old, 66-kg female client completes the 210-watts stage of the Storer cycling protocol. What is her VO$_2$max in ml·min^{-1}·kg^{-1}?

3. Your 22-year-old, 97-kg male client completes the 375-watts stage of the Storer cycling protocol. What is his VO$_2$max in ml·min^{-1}·kg^{-1}?

4. Your 67-year-old, 81-kg male client completes a 1-mile walk in 18 minutes and 14 seconds with a post-walk heart rate of 128 bpm. What is his VO$_2$max in ml·min^{-1}·kg^{-1}?

5. Your 47-year-old, 72-kg female client completes a 1-mile walk in 16 minutes and 30 seconds with a post-walk heart rate of 140 bpm. What is his VO$_2$max in ml·min^{-1}·kg^{-1}?

6. Your 32-year-old, 65-kg female client performs an Astrand bike test. Her average heart rate for the fifth and sixth minute at 900 kg·m·min^{-1} is 152 bpm. What is her VO$_2$max in ml·min^{-1}·kg^{-1}?

7. Your 21-year-old, 62-kg male client performs an Astrand bike test. His average heart rate for the fifth and sixth minute at 900 kg·m·min^{-1} is 156 bpm. What is his VO$_2$max in ml·min^{-1}·kg^{-1}?

8. Your 28-year-old, 87-kg client performs a submaximal bike test. His heart rates at the end of each stage are 110 bpm at 150 kg·m·min^{-1}, 138 bpm at 300 kg·m·min^{-1}, and 160 bpm at 450 kg·m·min^{-1}. What is his VO$_2$max in ml·min^{-1}·kg^{-1}?

9. Your 30-year-old, 59-kg client performs a submaximal bike test. Her heart rates at the end of each stage are 86 bpm at 150 kg·m·min^{-1}, 122 bpm at 600 kg·m·min^{-1}, 136 bpm at 750 kg·m·min^{-1}, and 152 bpm at 900 kg·m·min^{-1}. What is her VO$_2$max in ml·min^{-1}·kg^{-1}?

10. Your 72-year-old, 64-kg client performs a submaximal bike test. His heart rates at the end of each stage are 104 bpm at 150 kg·m·min^{-1} and 126 bpm at 300 kg·m·min^{-1}. What is his VO$_2$max in ml·min^{-1}·kg^{-1}?

Answers to Sample Problems in Calculating VO$_2$max

1. Answer: 26 ml·min^{-1}·kg^{-1}; 1.3 L·min^{-1}; 7.4 METs

   Solution: Just look it up on Table 7.1. To convert to L·min^{-1}, multiply by body weight and divide by 1000: 26 × 51/1000 = 1.326. To convert to METs, divide by 3.5: 26/3.5 = 7.429.

2. Answer: 36.8 ml·min^{-1}·kg^{-1}

   Solution: Set up the Storer equation for women and solve.

   VO$_2$max = [9.39(power) + 7.7(BW) − 5.88(age) + 136.7]/(BW)

   VO$_2$max = [9.39(210) + 7.7(66) − 5.88(32) + 136.7]/(66)

   VO$_2$max = [1971.9 + 508.2 − 188.16 + 136.7]/66

   VO$_2$max = 2428.64/66

   VO$_2$max = 36.798

3. Answer: 50.0 ml·min^{-1}·kg^{-1}

   Solution: Set up the Storer equation for men and solve.

   VO$_2$max = [10.51(power) + 6.35(BW) − 10.49(age) + 519.3]/(BW)

   VO$_2$max = [10.51(375) + 6.35(97) − 10.49(22) + 519.3]/(97)

   VO$_2$max = [3941.25 + 615.95 − 230.78 + 519.3]/97

   VO$_2$max = 4845.72/97

   VO$_2$max = 49.956

4. Answer: 23.5 ml·min^{-1}·kg^{-1}

   Solution: Set up the Rockport walking equation and solve. The time must be converted to minutes, i.e., 18 + 14/60 = 18.23 minutes.

   VO$_2$max = [6965.2 + 20.02(BW) − 25.7(age) + 595.5(sex) − 224(time) − 11.5(HR)]/(BW)

   VO$_2$max = [6965.2 + 20.02(81) − 25.7(67) + 595.5(1) − 224(18.23) − 11.5(128)]/(81)

   VO$_2$max = [6965.2 + 1621.62 − 1721.9 + 595.1 − 4083.52 − 1472]/81

VO$_2$max = 1904.5/81

VO$_2$max = 23.512

5. Answer: 26.3 ml·min^{-1}·kg^{-1}

Solution: Set up the Rockport walking equation and solve. The time must be converted to minutes, i.e., 16 + 30/60 = 16.50 minutes.

VO$_2$max = [6965.2 + 20.02(BW) − 25.7(age) + 595.5(sex) − 224(time) − 11.5(HR)]/(BW)

VO$_2$max = [6965.2 + 20.02(72) − 25.7(47) + 595.5(0) − 224(16.50) − 11.5(140)]/(72)

VO$_2$max = [6965.2 + 1441.44 − 1207.9 + 0 − 3696.0 − 1610]/72

VO$_2$max = 1892.74/72

VO$_2$max = 26.288

6. Answer: 46 ml·min^{-1}·kg^{-1}

Solution: From the Astrand nomogram, the raw VO$_2$max is approximately 3.3 L·min^{-1}, and the age correction factor is approximately 0.91. Thus, the absolute VO$_2$max is 3.3 × 0.91 = 3.003 L·min^{-1}; the relative VO$_2$max is 3.003 × 1000/65 = 46.20 ml·min^{-1}·kg^{-1}.

7. Answer: 49 ml·min^{-1}·kg^{-1}

Solution: From the Astrand nomogram, the raw VO$_2$max is approximately 2.9 L·min^{-1}, and the age correction factor is 1.04. Thus, the absolute VO$_2$max is 2.9 × 1.04 = 3.016 L·min^{-1}; the relative VO$_2$max is 3.016 × 1000/62 = 48.65 ml·min^{-1}·kg^{-1}.

8. Answer: 18 ml·min^{-1}·kg^{-1}

Solution: The age-predicted maximal heart rate is 220 − 28 = 192 bpm. From Figure 7.4, the estimated maximal workload at that heart rate is 640 kg·m·min^{-1}. Now we enter that into the cycle ergometry equation and solve for VO$_2$.

VO$_2$ = 3.5 + 2(640)/87

VO$_2$ = 3.5 + 1280/87

VO$_2$ = 3.5 + 14.71

VO$_2$ = 18.21

9. Answer: 47 ml·min^{-1}·kg^{-1}

Solution: The age-predicted maximal heart rate is 220 − 30 = 190 bpm. From Figure 7.4, the estimated maximal workload at that heart rate is 1280 kg·m·min^{-1}. Now we enter that into the cycle ergometry equation and solve for VO$_2$.

VO$_2$ = 3.5 + 2(1280)/59

VO$_2$ = 3.5 + 2560/59

VO$_2$ = 3.5 + 43.39

VO$_2$ = 46.89

10. Answer: 18 ml·min^{-1}·kg^{-1}

Solution: The age-predicted maximal heart rate is 220 − 72 = 148 bpm. From Figure 7.4, the estimated maximal workload at that heart rate is 450 kg·m·min^{-1}. Now we enter that into the cycle ergometry equation and solve for VO$_2$.

VO$_2$ = 3.5 + 2(450)/64

VO$_2$ = 3.5 + 900/64

VO$_2$ = 3.5 + 14.06

VO$_2$ = 17.56

Deriving the Simplified Equations

The metabolic equations of the American College of Sports Medicine (ACSM) have been simplified by changing the units for speed, grade, height, etc., into commonly used terms and then adjusting the conversion factors to compensate. Furthermore, when more than one conversion factor is used by the ACSM in the same term, these are combined to eliminate a mathematical step. Finally, all the simplified equations yield answers in the same oxygen consumption (VO_2) units, that is, $ml \cdot min^{-1} \cdot kg^{-1}$. Body mass is left in kilograms throughout the simplified equations, because the universally recognized units of VO_2 use kilograms. To convert a body mass in kilograms to body weight in pounds, simply multiply the kilograms by 2.2; alternatively, divide pounds by 2.2 to obtain kilograms (this is a rough approximation; a better approximation is 2.2046, but such a high degree of precision in the conversion factor is not needed when the original measurement is generally not recorded with a similar degree of accuracy).

It is important to emphasize that none of the simplifications change the mathematical content of the equations. They provide the same answers. The simplifications just make the equations easier to use.

WALKING EQUATION
The ACSM's version of the walking equation is as follows:

$$ml \cdot min^{-1} \cdot kg^{-1} = 3.5\ ml \cdot min^{-1} \cdot kg^{-1} + m/min \times 0.1 +$$
$$grade\ (frac) \times m/min \times 1.8$$

where:

1. The initial "$ml \cdot min^{-1} \cdot kg^{-1}$" is the gross VO_2, resting plus exercise, of the subject while walking. (Note: the ACSM places kg^{-1} as the middle term of the VO_2 units; for practical reasons it is placed at the end, i.e., $ml \cdot min^{-1} \cdot kg^{-1}$, throughout this book. There is no mathematical difference between the two versions.)

2. The "$3.5\ ml \cdot min^{-1} \cdot kg^{-1}$" represents resting VO_2.

3. The "$m/min \times 0.1$" is the VO_2 associated with the horizontal effort of walking; the value "0.1" is a conversion factor, technically equal to $0.1\ ml \cdot min^{-1} \cdot kg^{-1}$ of VO_2 for each $m \cdot min^{-1}$ of horizontal walking.

4. The "$grade\ (frac) \times m/min \times 1.8$" is the VO_2 associated with the vertical work of walking up a grade; the horizontal speed in $m \cdot min^{-1}$ multiplied by the fractional grade (i.e., the tangent of the angle the treadmill or road surface makes with the horizontal) yields the speed of vertical climbing. The value "1.8" is a conversion factor, technically equal to 1.8 $ml \cdot min^{-1} \cdot kg^{-1}$ of VO_2 for each $m \cdot min^{-1}$ of vertical climbing.

There are some limitations in the use of the equation. The resting component of VO_2, 3.5 $ml \cdot min^{-1} \cdot kg^{-1}$, is an average VO_2 for most persons. However, among individuals of the same body weight, those with less body fat have higher resting metabolisms. Also, there is some variation associated with the overall size of different persons, regardless of their body fat. Smaller individuals have slightly higher resting metabolisms. However, the value of 3.5 is a reasonably good approximation for most persons. The horizontal conversion factor of $0.1\ ml \cdot min^{-1} \cdot kg^{-1}$ of

VO_2 per m·min^{-1} of walking speed is an average value of the oxygen demand of walking within a narrow range of speeds. The conversion factor has the same value regardless of the age, sex, size, or fitness level of the subject. This is undoubtedly an oversimplification, but once again, it provides a fairly accurate approximation that can be used for a large population range. The ACSM states that the equation yields reasonably accurate values of VO_2 when the subject is walking between the speeds of 50 and 100 m·min^{-1}, or 1.9 and 3.7 mph. Thus, it is not appropriate to try to calculate the VO_2 of walking at substantially lower or higher speeds. However, as the long as the subject is truly walking, it is common to apply the equation to speeds somewhat outside this range. The vertical conversion factor of 1.8 ml·min^{-1}·kg^{-1} of VO_2 per m·min^{-1} of climbing is an average value of the oxygen demand of lifting one's body mass vertically. Walking and climbing (and running) are body-weight–dependent activities. The relative oxygen consumption (VO_2 expressed in kg^{-1} units) is very similar among different persons despite differences in body size.

If one measures speed in mph, it is necessary to convert to m·min^{-1} before using the equation. There are approximately 1609 meters in 1 mile and 60 minutes in 1 hour. Thus, to convert from mph to m·min^{-1}, one must perform the following manipulation:

$$\frac{miles}{hour} \times \frac{1609 \ meters}{1 \ mile} \times \frac{1 \ hour}{60 \ min} = \frac{meters}{min}$$

Thus, mph must be multiplied by $(1609/1) \times (1/60)$, or approximately 26.8. In the simplified equation, this is incorporated into the horizontal conversion factor, eliminating the need for this step:

$$(speed \times 0.1) \times 26.8 \ \frac{m \cdot min^{-1}}{mph} = speed \times (0.1 \times 26.8) = speed \times 2.68$$

Thus, the new term for the horizontal component in the walking equation becomes speed × 2.68, where speed is now in mph instead of m·min^{-1}.

Similarly, the speed in the vertical component is changed to mph by multiplying that conversion factor by 26.8, i.e., 1.8×26.8 = approximately 48. Before proceeding, however, another simplification is added at this point. The ACSM equation uses the fractional form for entering grade. For example, if a subject is walking on a treadmill set at a 12% grade, one enters 0.12 into the equation. Rather than dividing by 100 before entering the equation, let's just divide the conversion factor by 100. That way, grade remains in its percentage form at all times. Thus, the full conversion of the vertical component is as follows:

$$(grade \ (frac) \times speed \times 1.8) \times 26.8 \ \frac{m \cdot min^{-1}}{mph} \times \frac{1}{100\%} =$$
$$\% \ grade \times speed \times (1.8 \times 26.8/100) =$$
$$\% \ grade \times speed \times 0.48$$

Now, if the subject is walking up a 12% grade, just enter 12 into the equation. With all of the new conversion factors in place, the walking equation looks like this:

$$ml \cdot min^{-1} \cdot kg^{-1} = 3.5 \ ml \cdot min^{-1} \cdot kg^{-1} + speed \times 2.68 +$$
$$\% \ grade \times speed \times 0.48$$

Rearranging the terms to provide an equation that will be easier to work with yields the final version of the simplified walking equation:

WALKING

$$VO_2 = 3.5 + 2.68(speed) + 0.48(speed)(\% \ grade)$$
ml·min^{-1}·kg^{-1} mph mph

where it is understood that the VO_2 answer and the resting value of 3.5 are in ml·min^{-1}·kg^{-1}, whereas speed is in mph.

RUNNING EQUATION

The ACSM's treadmill running equation has the same format as the walking equation—a resting component, a horizontal component, and a vertical component—as follows:

$$\text{ml·min}^{-1}\text{·kg}^{-1} = 3.5\ \text{ml·min}^{-1}\text{·kg}^{-1} + \text{m/min} \times 0.2 +$$
$$\text{grade (frac)} \times \text{m/min} \times 0.9$$

where:

1. The initial "ml·min^{-1}·kg^{-1}" is the same as for the walking equation.

2. The "3.5 ml·min^{-1}·kg^{-1}" represents resting VO_2, just as in the walking equation.

3. The "m/min $\times$ 0.2" is again the VO_2 associated with the horizontal effort, this time of running. (Note that the conversion factor has been doubled, from 0.1 in the walking equation to 0.2; this is because running requires twice as much energy to cover the same distance as does walking—running is more energetic because it is basically jumping from one foot to the other.)

4. The "grade (frac) $\times$ m/min $\times$ 0.9" is again the VO_2 associated with the vertical work, this time of running up a grade. (Note that the conversion factor has been cut in half, from 1.8 for walking to 0.9 for running; the energy required to lift one's body vertically actually does not differ between running and walking. However, running on a treadmill is not the same as running outdoors; when jumping from one foot to the other while running on a treadmill, the treadmill belt is sliding beneath you, and you don't actually climb as much as you would if you always kept one foot in contact with the belt, as you do in walking. To account for this difference, the conversion factor has been reduced by half.)

As with the walking equation, there are limitations in the use of the running equation that should be considered. According to the ACSM, the equation is fairly accurate for speeds greater than 134 m·min^{-1}, i.e., greater than 5 mph. However, if the subject is truly jogging, during which both feet leave the ground between steps, then it can be used for speeds as low as 80 m·min^{-1}, or approximately 3 mph. The most important limitation in using the running equation is to remember that it is designed for running on a treadmill. If the subject is running outdoors up a hill, then the vertical component conversion factor must be doubled, so that it is equal to the conversion factor in the walking equation. If running down a hill, one cannot simply put a negative sign on the grade and subtract the vertical VO_2 component. Although the subject is gaining physical work from gravity, he or she is also performing physiologic work in the eccentric muscle actions needed to slow the descent. Until sufficient research is performed to determine how much oxygen consumption is occurring under these circumstances, it is best to avoid using the running equation for downhill running. Of course, if the outdoor running course is flat, then the grade is zero, and the vertical component drops out of the equation.

As done previously for the walking equation, the horizontal and vertical conversion factors are changed in the simplified version of the equation to account for the use of mph for speed and percentage units for grade. Because the changes are exactly as described above, they will not be repeated here in detail. Rather, one sees that the horizontal conversion factor in the simplified running equation is twice that in the simplified walking equation (i.e., 5.36 versus 2.68), and the vertical conversion factor is half that in the simplified walking equation (i.e., 0.24 versus 0.48). The final version of the simplified treadmill running equation is:

TREADMILL RUNNING

$$VO_2 = 3.5 + 5.36(\text{speed}) + 0.24(\text{speed})(\%\ \text{grade})$$
$$\text{ml·min}^{-1}\text{·kg}^{-1} \qquad \text{mph} \qquad\qquad \text{mph}$$

where it is understood that the VO_2 answer and the resting value of 3.5 are in $ml\cdot min^{-1}\cdot kg^{-1}$, whereas speed is in mph.

For running uphill while outdoors, the equation takes the following form:

OUTDOOR
RUNNING

$$VO_2 = 3.5 + 5.36(speed) + 0.48(speed)(\% \text{ grade})$$
$$ml\cdot min^{-1}\cdot kg^{-1} \qquad mph \qquad\qquad mph$$

where again VO_2 is in $ml\cdot min^{-1}\cdot kg^{-1}$, and speed is in mph.

CYCLE ERGOMETRY EQUATIONS

The ACSM's equations for cycling exercise on a leg or arm ergometer have a very different format than do the walking and running equations. To simplify these ergometry equations, they are rearranged into the same format as for walking and running.

The ACSM's leg ergometer equation is as follows:

$$ml/min = 3.5\ ml\cdot min^{-1}\cdot kg^{-1} \times kg\ BW + None + kgm/min \times 2$$

where:

1. The initial "ml/min" is the gross VO_2, resting plus exercise, for cycling on a stationary ergometer. Note that this is the absolute amount of oxygen used per minute (in ml/min, i.e., $ml\cdot min^{-1}$), not the relative amount used for each kilogram of body weight.

2. The "$3.5\ ml\cdot min^{-1}\cdot kg^{-1} \times kg\ BW$" is resting VO_2 for the total body in $ml\cdot min^{-1}$, obtained by multiplying the relative VO_2 of $3.5\ ml\cdot min^{-1}\cdot kg^{-1}$ by the subject's body weight (BW) in kilograms.

3. The "none" simply means that there is not a horizontal component in the ergometry equation, as there was in the walking and running equations.

4. The "$kgm/min \times 2$" is the VO_2 associated with the vertical or, actually, the resistive work of pedaling against the ergometer; the power level of the subject on the ergometer is given in kgm/min (i.e., $kg\cdot m\cdot min^{-1}$). These are shorthand units sometimes used by exercise scientists to indicate power; they are obtained by multiplying the resistance setting (in kilograms) on the ergometer times the distance the flywheel travels for each revolution of the pedals (6 m on a Monark bicycle, 3 m on a Tunturi or BodyGuard) times the pedaling rate in rpm. The "$\times 2$" is a conversion factor, technically equal to $2\ ml\cdot min^{-1}$ of VO_2 for each $kg\cdot m\cdot min^{-1}$ of power. More simply, it takes 2 ml of oxygen to push the flywheel on the bike 1 meter against a strap that has a resistance equivalent to 1 kg (kilograms are not actually units of resistive force, but the force is measured by comparing it to the force of gravity on 1 kg). This 2 ml of oxygen is based on the 1.8 ml of oxygen needed for raising 1 kg 1 meter in the vertical component of the walking equation and is increased here by 0.2, to 2.0, to account for additional friction within the drive train of the ergometer.

The primary limitation of the leg ergometry equation is that the conversion factor of 2 has not been well established. According to the ACSM, the equation is accurate for power levels between 300 and 1200 $kg\cdot m\cdot min^{-1}$. However, the equation is routinely used for any power level that can be maintained under steady-state conditions. In fact, it is even used to estimate the VO_2 during maximal aerobic conditions, as described in Chapter 7.

Having the ergometry equations give VO_2 in $ml\cdot min^{-1}$, when the other equations use $ml\cdot min^{-1}\cdot kg^{-1}$, is awkward and at times confusing. To simplify the equations, it is only necessary to divide both sides by the subject's body mass in kilograms:

$$\frac{ml/min}{kg\ BW} = \frac{3.5\ ml \cdot min^{-1} \cdot kg^{-1} \times kg\ BW}{kg\ BW} + \frac{None}{kg\ BW} + \frac{kgm/min \times 2}{kg\ BW}$$

Dividing by "kg BW" causes that to drop out of the resting VO$_2$ term. "None" divided by anything is zero. This yields:

$$ml \cdot min^{-1} \cdot kg^{-1} = 3.5\ ml \cdot min^{-1} \cdot kg^{-1} + 0 + (kgm/min \times 2)/(kg\ BW)$$

Now, if we remove the zero and simplify the remaining terms, the final version of our simplified leg ergometry equation is as follows:

LEG ERGOMETRY

VO$_2$ = 3.5 + 2(workload)/BW
ml·min^{-1}·kg^{-1} kg·m·min^{-1} kg

where it is understood that the VO$_2$ answer and the resting value of 3.5 are in ml·min^{-1}·kg^{-1}, workload (i.e., power) is in kg·m·min^{-1}, and body weight is in kg.

The arm ergometry equation is nearly identical. The only difference is that the conversion factor of 2 ml·min^{-1} of VO$_2$ for each kg·m·min$^-$ of power is raised to 3. This increase is based on a greater oxygen cost when a given amount of work is being done by a small muscle mass. In other words, it is less efficient to do work with the arms instead of the legs. Because this conversion factor is the only difference between the leg and arm equations, the simplified arm ergometry equation is just:

ARM ERGOMETRY

VO$_2$ = 3.5 + 3(workload)/BW
ml·min^{-1}·kg^{-1} kg·m·min^{-1} kg

as in the leg equation, it is understood that the VO$_2$ answer and the resting value of 3.5 are in ml·min^{-1}·kg^{-1}, workload (i.e., power) is in kg·m·min^{-1}, and body weight is in kilograms.

BENCH STEPPING EQUATION
The ACSM's bench stepping equation is as follows:

$$ml \cdot min^{-1} \cdot kg^{-1} = (incl.\ in\ horiz.\ and\ vert.) + steps/min \times 0.35 +$$
$$m/step \times steps/min \times 1.33 \times 1.8$$

where:

1. The initial "ml·min^{-1}·kg^{-1}" is the gross VO$_2$, back in the per kg units.

2. The "incl. in horiz. and vert." is a statement indicating that the resting VO$_2$ of 3.5 ml·min^{-1}·kg^{-1} will not be part of the equation, because it is "included" in the VO$_2$ determination from the horizontal and vertical components of the equation (actually, it is included only in the horizontal component). This is unfortunate, because it means the equation has a different format than the previous ones.

3. The "steps/min × 0.35" is the VO$_2$ associated with the "horizontal" aspect of stepping up and down; the "0.35" is a conversion factor, technically equal to 0.35 ml·kg^{-1} of oxygen for each step the subject takes.

4. The "m/step × steps/min × 1.33 × 1.8" is the VO$_2$ associated with the vertical work of stepping; "m/step × steps/min" yields the vertical distance that the subject climbs each

minute (in which m/step is simply the step height in meters). The 1.8 conversion factor is the same one seen in the uphill walking conversion, i.e., 1.8 ml·min^{-1}·kg^{-1} for each m·min^{-1} of climbing. The 1.33 is an adjustment for the eccentric portion of the stepping, that is to say, you use 1.8 units of oxygen in stepping up and another third on top of that in stepping back down. More technically, the vertical component could have been written as (m·step^{-1} × steps·min^{-1} × 1.8) × 1.33.

The major limitation to the bench stepping equation is the assumption that the eccentric phase (stepping down) requires one third as much work as the concentric phase (stepping up). The ACSM states that the equation is accurate for power outputs ranging from 300 to 1200 kg·m·min^{-1}. If you wished to determine the power output, first calculate the vertical work as: (step height in meters) × (steps·min^{-1}) × (kg of BW). Then multiply this value by 1.33 to add the eccentric component. Under practical situations, it is unlikely that one would perform this calculation first in an attempt to determine if the workload is between 300 and 1200 kg·m·min^{-1}. Rather, most users should assume, as in leg ergometry, that the equation gives reasonably accurate VO$_2$s as long as the subject is exercising under steady-state conditions.

To simplify this equation, let's multiply the 1.8 × 1.33 to produce a single conversion factor in the vertical component. Next, because most users will measure bench height in inches, let's divide the conversion factor by the number of inches in 1 meter, i.e., 39.37. Let's also substitute a zero for the "included in horizontal and vertical" statement regarding the resting component:

ml·min^{-1}·kg^{-1} = 0 + steps/min × 0.35 + (m/step × steps/min × 1.33 × 1.8)/39.37

= steps/min × 0.35 + (height × steps/min) × 2.4/39.37

= steps/min × 0.35 + (height × steps/min) × 0.061

Rearranging the terms provides the final version of the simplified bench stepping equation:

STEPPING

$$VO_2 = 0.35(\text{rate}) + 0.061(\text{rate})(\text{height})$$

ml·min^{-1}·kg^{-1} steps·min^{-1} inches

where it is understood that the VO$_2$ answer is in ml·min^{-1}·kg^{-1}, rate is in steps·min^{-1}, and height is in inches.

Reference

1. American College of Sports Medicine. ACSM's guidelines for exercise testing and prescription. 5th ed. Baltimore: Williams & Wilkins, 1995:278–283.

Case Studies

These case studies give the reader the chance to put all the information they have learned together. Here is how the calculations described in this book are used when dealing with real clients.

CASE 1

Bob is a 51-year-old bank manager. He is 5 feet, 8 inches tall and weighs 165 pounds. He has a resting HR of 78 bpm. You perform a submaximal bike test on Bob and obtain the following HR data: 100 bpm at 150 kg·m·min^{-1}, 120 bpm at 300 kg·m·min^{-1}, and 138 at 450 kg·m·min^{-1}. You attempted 600 kg·m·min^{-1}, but his HR reached 148 bpm at the end of the first minute, and you terminated the test because this exceeded 85% of age-predicted maximum HR.

What is Bob's estimated VO$_2$max? Using the percentage HRR method, what would be his target HR prescription at 60 to 70% of VO$_2$max? Bob feels most comfortable pedaling a Monark bike at 60 rpm. What workload and kg setting should you prescribe for him to be at 60 to 70% of VO$_2$max?

VO$_2$max

Bob has three steady-state HRs from his submaximal bike test; the 148 bpm in stage IV should be ignored. His estimated maximal HR is $220 - 51 = 169$ bpm. Plotting the data from his first three stages and extrapolating to 169 bpm yields an estimated maximal workload of 690 kg·m·min^{-1}. Use the leg ergometry formula to solve for VO$_2$max. Note that his body mass is $165/2.2 = 75$ kg.

$$VO_2 = 3.5 + 2(690)/75$$
$$VO_2 = 3.5 + 1380/75$$
$$VO_2 = 3.5 + 18.4$$
$$VO_2 = 21.9$$

His VO$_2$max is estimated to be 21.9 ml·min^{-1}·kg^{-1}.

HR Prescription

$$\text{lower target HR} = 0.60(169 - 78) + 78$$
$$= 0.60(91) + 78$$
$$= 54.6 + 78$$
$$= 132.6$$
$$\text{upper target HR} = 0.70(91) + 78$$
$$= 63.7 + 78$$
$$= 141.7$$

His HR prescription is 133 to 142 bpm.

Workload Prescription

The desired exercise VO$_2$ is from 60 to 70% of VO$_2$max, i.e., from $0.60 \times 21.9 = 13.1$ to $0.70 \times 21.9 = 15.3$ ml·min^{-1}·kg^{-1}. The workloads for these VO$_2$s can be calculated from the leg ergometry formula:

lower VO$_2$:
$$13.1 = 3.5 + 2(\text{workload})/75$$
$$13.1 - 3.5 = 2(\text{workload})/75$$
$$9.6 = 2(\text{workload})/75$$
$$9.6 \times 75 = 2(\text{workload})$$

$$720 = 2(\text{workload})$$
$$720/2 = \text{workload}$$
$$360 = \text{workload}$$

upper VO_2: $15.3 = 3.5 + 2(\text{workload})/75$
$$15.3 - 3.5 = 2(\text{workload})/75$$
$$11.8 = 2(\text{workload})/75$$
$$11.8 \times 75 = 2(\text{workload})$$
$$885 = 2(\text{workload})$$
$$885/2 = \text{workload}$$
$$442.5 = \text{workload}$$

The target workload is from approximately 360 to 443 kg·m·min^{-1}. Determine the kg settings for each at 60 rpm as follows:

lower workload: $360 = (\text{kg setting})(6)(60)$
$$360 = (\text{kg setting})(360)$$
$$360/360 = \text{kg setting}$$
$$1.00 = \text{kg setting}$$

upper workload: $443 = (\text{kg setting})(360)$
$$443/360 = \text{kg setting}$$
$$1.23 = \text{kg setting}$$

The Monark resistance should be set at 1 to 1¼ kg.

CASE 2

Karen is a 27-year-old physical therapist and recreational runner. She is 5 feet, 4 inches tall and weighs 124 pounds. Her resting HR is 62 bpm. You perform a maximal Bruce protocol test on her, and she stops at 16 minutes, 23 seconds with a HR of 203 bpm.

What is her estimated VO_2max? Using the percentage HRR method, what would be her target HR prescription at 75 to 85% of VO_2max? If she wishes to perform aerobic interval training at VO_2max, at what speed would that be on flat ground?

VO₂max

According to Table 7.1, her time of 16 minutes, 23 seconds translates to a VO_2max of 60 ml·min^{-1}·kg^{-1}.

HR Prescription

Her known maximal HR is 203 bpm. Do not use $220 - 27$.

lower target HR $= 0.75(203 - 62) + 62$
$$= 0.75(141) + 62$$
$$= 105.75 + 62$$
$$= 167.75$$
upper target HR $= 0.85(141) + 62$
$$= 119.85 + 62$$
$$= 181.85$$

Her HR prescription is 168 to 182 bpm.

Speed at VO₂max

Use the running equation and solve for speed. Because she is on flat ground, the vertical component drops out.

$$60 = 3.5 + 5.36(\text{speed})$$
$$60 - 3.5 = 5.36(\text{speed})$$
$$56.5 = 5.36(\text{speed})$$
$$56.5/5.36 = \text{speed}$$
$$10.54 = \text{speed}$$

She should run her aerobic intervals at approximately 10.5 mph. This translates to a pace of 60/10.5 = 5.71 min/mile, or 5 min and 0.71 × 60 sec = 5 min and 43 sec.

CASE 3

Frank is a 62-year-old patient with cardiac disease referred to your Phase III rehabilitation program. Four months ago, after complaints of chest pain, angiography revealed a distal occlusion of the left circumflex coronary artery. He underwent a successful percutaneous transluminal coronary angioplasty. He has recently completed 3 months of Phase II rehabilitation without restenosis or chest pain.

At the completion of Phase II, he underwent a thallium scan Bruce treadmill test. His resting HR was 65 bpm, with blood pressure of 110/92 mm Hg. His maximal HR was 139 bpm, with blood pressure of 155/98. He completed the third stage of the protocol (3.4 mph and 14% grade).

Based on the speed and grade during his exercise test, what is Frank's estimated VO_2max? What is his target HR prescription at 60 to 80% of VO_2max? Frank is comfortable walking on a treadmill at 2.5 mph. What grade should you prescribe for him to be at 60 to 80% of VO_2max?

VO_2max

$$VO_2 = 3.5 + 2.68(3.4) + 0.48(3.4)(14)$$
$$VO_2 = 3.5 + 9.112 + 22.848$$
$$VO_2 = 35.46$$

His VO_2max is approximately 35.5 $ml\cdot min^{-1}\cdot kg^{-1}$.

HR Prescription

$$\begin{aligned} lower\,target\,HR &= 0.60(139 - 65) + 65 \\ &= 0.60(74) + 65 \\ &= 44.4 + 65 \\ &= 109.4 \\ upper\,target\,HR &= 0.80(74) + 65 \\ &= 59.2 + 65 \\ &= 124.2 \end{aligned}$$

His HR prescription is from 109 to 124 bpm.

Workload Prescription

The desired exercise VO_2 is from 60 to 80% of VO_2max, i.e., from 0.60 × 35.5 = 21.3 to 0.80 × 35.5 = 28.4 $ml\cdot min^{-1}\cdot kg^{-1}$. The treadmill grades at 2.5 mph for these VO_2s can be calculated from the walking equation:

lower VO_2:
$$21.3 = 3.5 + 2.68(2.5) + 0.48(2.5)(\%\,grade)$$
$$21.3 = 3.5 + 6.7 + 1.2(\%\,grade)$$
$$21.3 = 10.2 + 1.2(\%\,grade)$$
$$21.3 - 10.2 = 1.2(\%\,grade)$$
$$11.1 = 1.2(\%\,grade)$$
$$11.1/1.2 = \%\,grade$$
$$9.25 = \%\,grade$$

upper VO_2:
$$28.4 = 10.2 + 1.2(\%\,grade)$$
$$28.4 - 10.2 = 1.2(\%\,grade)$$
$$18.2 = 1.2(\%\,grade)$$
$$18.2/1.2 = \%\,grade$$
$$15.17 = \%\,grade$$

The treadmill grade should be set from 9.3 to 15.2%.

Additional Sample Problems

Note: Questions marked by an asterisk are based on supplemental material that goes beyond the scope of the American College of Sports Medicine guidelines.

ADDITIONAL SAMPLE PROBLEMS

1. What would be the VO_2 in $ml \cdot min^{-1} \cdot kg^{-1}$ for a 80-kg man to walk at 3.0 mph and 5% grade?

2. What would be the approximate caloric energy expenditure of a 90-kg person walking 2.5 mph on a treadmill up a 6% grade for 30 minutes?

3. What would be the MET level for a 85-kg person walking on a treadmill at 3.7 mph and 5% grade?

4. What is the VO_2 requirement of treadmill walking at 2.8 mph, 6% grade?

5. What is the energy requirement in METs for a client who is walking 3.9 mph?

6. A patient with cardiac disease cannot exceed 2 mph on the treadmill but needs to exercise at 8.0 METs. What percentage grade do you set?

7. A client cannot exercise above a MET level of 5.2. You must prescribe a walking program assuming 5% grade. At what speed would the client walk?

8. Your 150-lb client is walking on a treadmill at 3.0 mph up a 3% grade for 30 minutes, 3 times per week. Assuming a constant caloric intake, how long must your client exercise to lose 10 lb of fat with this exercise prescription?

9. Your 68-kg client is walking on a treadmill at 3.7 mph up a 10% grade. What is the total amount of oxygen in liters that your client will consume over 1 hour?

10. Your 48-kg client has a VO_2max of 2.4 $L \cdot min^{-1}$. You want him to exercise to 75% of his VO_2max. Assuming the treadmill is set at a 12% grade, what walking speed would you set?

11. Calculate the VO_2 in $L \cdot min^{-1}$ for a 90-kg client who runs on a treadmill at 7 mph up a 3% incline.

12. Running on a treadmill at 6 mph and 10% grade requires what MET level?

13. Running outdoors at 9 mph elicits what VO_2 in $ml \cdot min^{-1} \cdot kg^{-1}$?

14. What is the MET cost of running on a treadmill at 5.2 mph and 15% grade?

15. Your client weighs 250 lb. You want him to burn 0.5 lb of fat each week from exercise. He plans to run outdoors 4 times a week at 9 mph. How long should he run each day?

16. Your client has a functional capacity of 11 METs. You want him to train at 80% of his functional capacity. He will be running on a treadmill at 4 mph. What grade should be used?

17. A 60-kg woman is running on a treadmill at 10 mph and 5% grade. What is her energy expenditure in $kcal \cdot min^{-1}$?

18. A friend of yours has finished a 5-mile race in 30 minutes and 18 seconds. What was his VO_2 during the race in $ml \cdot min^{-1} \cdot kg^{-1}$?

19. If your friend in the previous question had to climb a 6% grade hill outdoors, how much would your friend have to slow down? (Assume the same VO_2)

20. Your client has a VO_2max of 70 ml·min^{-1}·kg^{-1}. You want him to exercise at 85% of his VO_2max on a treadmill. If the speed is 7.5 mph, what percentage grade should be used?

21. A 90-kg client is exercising on a stationary bike at 1500 kg·m·min^{-1}. What is his estimated VO_2 in ml·min^{-1}·kg^{-1}?

22. A 75-kg friend of yours is exercising on a stationary bike at 900 kg·min^{-1}·kg^{-1}. How many kilocalories would he expend over 60 minutes?

23. A 70-kg client is exercising on a stationary bike at 200 watts. What is his estimated VO_2 in ml·min^{-1}·kg^{-1}?

24. What workload in kg·m·min^{-1} should be prescribed if you want your 91-kg client to exercise at 6.0 kcal·min^{-1} on a stationary bike?

25. You want your 110-kg friend to exercise at 75% of his 5.0 L·min^{-1} capacity on a Monark bike. If he pedals at 80 rpm, what resistance setting is needed?

26. You want your 70-kg client to exercise at 12 METs on a Tunturi bike. If she pedals at 80 rpm, what resistance setting is needed?

27. A 90-kg individual is arm cranking on a Monark leg ergometer at 800 kg·m·min^{-1}. What is the VO_2 in ml·min^{-1}·kg^{-1}?

28. Consider the same individual described above, only this time the workload is 50 watts. What is the VO_2 in L·min^{-1}?

29. A 50-kg cardiopulmonary patient is arm cranking on a Monark arm ergometer at 50 rpm with a resistance setting of 0.5 kg. What is the VO_2 in ml·min^{-1}·kg^{-1}?

30. You want your 70-kg client to burn 3500 kcal per week on a stationary bike. If she exercises at 1000 kg·m·min^{-1}, how many hours per week must she exercise?

31. What is the VO_2 in ml·min^{-1}·kg^{-1} for stepping on a 12-inch step at a rate of 20 steps per minute?

32. What is the VO_2 in liters per minute for your 90-kg client who uses a 20-inch step at a rate of 35 steps per minute?

33. Your 70-kg friend is stepping up a 10-inch step at 40 steps per minute. What is her energy expenditure in kcal·min^{-1}?

34. You teach a step aerobics class that uses 3-inch stepping platforms that can be added on top of each other. Your music has a beat set at 140 bpm. You plan to have your class step 35 steps per minute. Assume that the average individual MET capacity is 11 METs. How many step platforms should they use to exercise at 85% of their capacity?

35. You want your 90-kg client to exercise at 1.6 L·min^{-1} while stepping on a 15-inch bench. At what rate would you set the metronome?

*36. Your client is exercising at level 8 on a Stairmaster. What is the VO_2 in ml·min^{-1}·kg^{-1}?

*37. You want your client to exercise at 70% of his 50 ml·min^{-1}·kg^{-1} capacity. What setting on the Stairmaster would you use?

*38. Your 70-kg client exercised for 30 minutes on a Concept II rower at an average power of 150 watts. How many kilocalories did he burn?

*39. Your 90-kg client has a peak VO_2 during arm exercise of 30 ml·min^{-1}·kg^{-1}. At what level on a Schwinn Airdyne should he exercise to achieve 75% of this?

*40. You want your 60-kg client to exercise at 8 METs. At what speed should he ride his bicycle outdoors?

41. You have a 30-year-old client who wants to exercise at 75% of his functional capacity. His resting HR is 70 bpm. What is his target HR using the percentage HRR method?

42. Your 80-year-old client has a maximum HR of 120 bpm and a resting HR of 82 bpm. What would be his target HR at 45% of VO_2max using the percentage HRR method?

43. Your 20-year-old friend has a resting HR of 60 bpm. What would be her target HR at 85% of VO_2max using the percentage HRR method?

*44. You want your 70-year-old client to exercise at 70% of VO_2max. Using Table 6.1 and the percentage HRmax method, what would be his target HR?

*45. During an aerobic training session, how many beats in 10 seconds should your client obtain if he wants to maintain his workout at 80% of VO_2max? Your client is 22 years old. Use Table 6.1 and the percentage HRmax method.

*46. Your 70-kg client completes 9 minutes and 15 seconds on the Bruce protocol treadmill test. What is the VO_2max in $L \cdot min^{-1}$?

*47. A 45-year-old, 90-kg man completes the 195-watts stage of the Storer cycling protocol. What is his VO_2max in $ml \cdot min^{-1} \cdot kg^{-1}$?

*48. Your 52-year-old, 75-kg male client completes a 1 mile walk in 20 minutes and 17 seconds, with a post-walk HR of 130 bpm. What is his VO_2max in $ml \cdot min^{-1} \cdot kg^{-1}$?

49. Your 40-year-old, 70-kg male client performs an Astrand bike test. His average HR for the fifth and sixth minute at 750 $kg \cdot m \cdot min^{-1}$ is 160 bpm. What is his VO_2max in $ml \cdot min^{-1} \cdot kg^{-1}$?

50. Your 30-year-old, 90-kg client performs a submaximal bike test. His HRs at the end of each stage are 115 bpm at 150 $kg \cdot m \cdot min^{-1}$, 130 at 300 $kg \cdot m \cdot min^{-1}$, and 150 at 450 $kg \cdot m \cdot min^{-1}$. What is his VO_2max in $ml \cdot min^{-1} \cdot kg^{-1}$?

ANSWERS TO ADDITIONAL SAMPLE PROBLEMS

1. Answer: 18.7 $ml \cdot min^{-1} \cdot kg^{-1}$

   Solution:

   $VO_2 = 3.5 + 2.68(3.0) + 0.48(3.0)(5)$

   $= 3.5 + 8.04 + 7.2$

   $= 18.74$

2. Answer: 235 kcal

   Solution:

   $VO_2 = 3.5 + 2.68(2.5) + 0.48(2.5)(6)$

   $= 3.5 + 6.7 + 7.2$

   $= 17.4$ $ml \cdot min^{-1} \cdot kg^{-1}$

   Convert to $L \cdot min^{-1}$: $(17.4 \times 90$ kg$)/1000 = 1.566$

   Convert to kcal: 1.566×5 $kcal \cdot L^{-1} \times 30$ min $= 235$

3. Answer: 6.4 METs

   Solution:

   $VO_2 = 3.5 + 2.68(3.7) + 0.48(3.7)(5)$

   $= 3.5 + 9.916 + 8.88$

   $= 22.296$ $ml \cdot min^{-1} \cdot kg^{-1}$

   METs $= 22.296/3.5 = 6.4$

4. Answer: 19.1 ml·min^{-1}·kg^{-1}

Solution:

VO_2 = 3.5 + 2.68(2.8) + 0.48(2.8)(6)

      = 3.5 + 7.504 + 8.064

      = 19.068

5. Answer: 4.0 METs

Solution:

VO_2 = 3.5 + 2.68(3.9) + 0.40(3.9)(0)

      = 3.5 + 10.452 + 0

      = 13.952 ml·min^{-1}·kg^{-1}

Convert to METs: 13.952/3.5 = 3.986

6. Answer: 20%

Solution:

8.0 METs × 3.5 = 28 ml·min^{-1}·kg^{-1}

28 = 3.5 + 2.68(2.0) + 0.48(2.0)(%grade)

28 = 3.5 + 5.36 + 0.96(%grade)

19.4 = 0.96(%grade)

19.14/0.96 = %grade

19.9 = %grade

7. Answer: 2.9 mph

Solution:

5.2 METs × 3.5 = 18.2 ml·min^{-1}·kg^{-1}

18.2 = 3.5 + 2.68(speed) + 0.48(speed)(5)

18.2 = 3.5 + 2.68(speed) + 2.4(speed)

14.7 = (2.68 + 2.4)(speed)

14.7 = 5.08(speed)

14.7/5.08 = speed

2.89 = speed

8. Answer: 72 weeks

Solution:

VO_2 = 3.5 + 2.68(3) + 0.48(3)(3)

      = 3.5 + 8.04 + 4.32

      = 15.86

Convert to L·min^{-1}: (15.86 × 68 kg)/1000 = 1.08

Convert to kcal·min^{-1}: 1.08 L·min^{-1} × 5 kcal·L^{-1} = 5.4 kcal·min^{-1}

5.4 × 30 min × 3 times per week = 486 kcal each week

1 lb fat = 3500 kcal

10 lb fat = 35,000 kcal

35,000 kcal/486 kcal per week = 72 weeks

9. Answer: 126 liters

Solution:

$VO_2 = 3.5 + 2.68(3.7) + 0.48(3.7)(10)$

$= 3.5 + 9.916 + 17.76$

$= 31.2 \text{ ml} \cdot \text{min}^{-1} \cdot \text{kg}^{-1}$

Convert to $L \cdot \text{min}^{-1}$: $(31.1 \times 68)/1000 = 2.1 \text{ L} \cdot \text{min}^{-1}$

$2.1 \text{ L} \cdot \text{min}^{-1} \times 60 \text{ minutes} = 126$

10. Answer: 4.0 mph

Solution:

$VO_2\text{max in ml} \cdot \text{min}^{-1} \cdot \text{kg}^{-1} = (2.4 \times 1000)/48 = 50$

75% of 50 = $0.75 \times 50 = 37.5 \text{ ml} \cdot \text{min}^{-1} \cdot \text{kg}^{-1}$

$37.5 = 3.5 + 2.68(\text{speed}) + 0.48(\text{speed})(12)$

$34 = 2.68(\text{speed}) + 5.76(\text{speed})$

$34 = 8.44(\text{speed})$

$34/8.44 = \text{speed}$

$4.028 = \text{speed}$

11. Answer: 4.1 $L \cdot \text{min}^{-1}$

Solution:

$VO_2 = 3.5 + 5.36(7) + 0.24(7)(3)$

$= 3.5 + 37.52 + 5.04$

$= 46.06 \text{ ml} \cdot \text{min}^{-1} \cdot \text{kg}^{-1}$

Convert to $L \cdot \text{min}^{-1}$: $(46.06 \times 90 \text{ kg})/1000 = 4.1$

12. Answer: 14.3 METs

Solution:

$VO_2 = 3.5 + 5.36(6) + 0.24(6)(10)$

$= 3.5 + 32.16 + 14.4$

$= 50.06 \text{ ml} \cdot \text{min}^{-1} \cdot \text{kg}^{-1}$

Convert to METs: $50.06/3.5 = 14.3$

13. Answer: 51.7 $\text{ml} \cdot \text{min}^{-1} \cdot \text{kg}^{-1}$

Solution:

$VO_2 = 3.5 + 5.36(9) + 0.48(9)(0)$

$= 3.5 + 48.24 + 0$

$= 51.74$

14. Answer: 14.3 METs

Solution:

$VO_2 = 3.5 + 5.36(5.2) + 0.24(5.2)(15)$

$= 3.5 + 27.872 + 18.72$

$= 50.09$ ml·min^{-1}·kg^{-1}

Convert to METs: $50.09/3.5 = 14.3$

15. Answer: 15 minutes

Solution:

$VO_2 = 3.5 + 5.36(9) + 0.48(9)(0)$

$= 3.5 + 48.24 + 0$

$= 51.74$ ml·min^{-1}·kg^{-1}

0.5 lb of fat $= 3500/2 = 1750$ kcal

1750/4 days per week $= 437.5$ kcal per exercise session

Convert VO_2 to L·min^{-1}: $(51.74 \times 113$ kg$)/1000 = 5.8$

Convert to kcal·min^{-1}: 5.8×5 kcal·L$^{-1} = 29.2$ kcal·min^{-1}

Time needed is: $437.5/29.2 = 14.9$ minutes

16. Answer: 6.1% grade

Solution:

80% of 11 METs is 8.8 METs

$VO_2 = 8.8 \times 3.5 = 30.8$ ml·min^{-1}·kg^{-1}

$30.8 = 3.5 + 5.36(4) + 0.24(4)(\%grade)$

$30.8 = 3.5 + 21.44 + 0.96(\%grade)$

$30.8 = 24.94 + 0.96 (\%grade)$

$5.86 = 0.96(\%grade)$

$5.86/0.96 = \%grade$

$6.1 = \%grade$

17. Answer: 20.7 kcal·min^{-1}

Solution:

$VO_2 = 3.5 + 5.36(10) + 0.24(10)(5)$

$= 3.5 + 53.6 + 12.0$

$= 69.1$ ml·min^{-1}·kg^{-1}

Convert to L·min^{-1}: $(69.1 \times 60$ kg$)/1000 = 4.146$ L·min^{-1}

Convert to kcal·min^{-1}: 4.146×5 kcal·L$^{-1} = 20.73$

18. Answer: 57 ml·min^{-1}·kg^{-1}

Solution:

Convert seconds to minutes: $18/60 = 0.30$

Therefore, 30:18 seconds $= 30.30$ minutes

Convert minutes to hours: $30.30/60 = 0.505$ hours

Speed $= 5/0.505 = 9.9$ mph

$VO_2 = 3.5 + 5.36(9.9) + 0.48(9.9)(0)$

$= 3.5 + 53.064 + 0$

$= 56.564$

19. Answer: Slow from 9.9 to 6.5 mph

    Solution:

    57 = 3.5 + 5.36(speed) + 0.48(speed)(6)

    57 = 3.5 + 5.36(speed) + 2.88(speed)

    57 = 3.5 + 8.24(speed)

    53.5 = 8.24(speed)

    53.5/8.24 = speed

    6.49 = speed

20. Answer: 8.8% grade

    Solution:

    85% of 70 ml·min^{-1}·kg^{-1} = 59.5 ml·min^{-1}·kg^{-1}

    59.5 = 3.5 + 5.36(7.5) + 0.24(7.5)(%grade)

    59.5 = 3.5 + 40.2 + 1.8(%grade)

    15.8 = 1.8(%grade)

    15.8/1.8 = %grade

    8.777 = %grade

21. Answer: 36.8 ml·min^{-1}·kg^{-1}

    Solution:

    VO_2 = 3.5 + 2(1500)/90

         = 3.5 + 33.33

         = 36.83

22. Answer: 619 kcal

    Solution:

    VO_2 = 3.5 + 2(900)/75

         = 3.5 + 24

         = 27.5

    Convert to L·min^{-1}: (27.5 × 75 kg)/75 = 2.0625

    Convert to kcal: 2.0625 × 5 kcal·min^{-1} = 10.3125

    10.3125 × 60 minutes = 618.75

23. Answer: 37.8 ml·min^{-1}·kg^{-1}

    Solution:

    200 watts × 6 = 1200 kg·m·min^{-1}

    VO_2 = 3.5 + 2(1200)/70

         = 3.5 + 34.29

         = 37.79

24. Answer: 441 kg·m·min^{-1}

    Solution:

    6.0 kcal·min^{-1}/5 kcal·L^{-1} = 1.2 L·min^{-1}

Convert VO$_2$ to ml·min^{-1}·kg^{-1}: (1.2 × 1000)/91 kg = 13.2

13.2 = 3.5 + 2(workload)/91

9.7 = 2(workload)91

9.7/2 = workload/91

4.85 × 91 = workload

441.35 = workload

25. Answer: 3.5 kg

Solution:

5.0 L·min^{-1} × 0.75 = 3.75 L·min^{-1}

Convert VO$_2$ to ml·min^{-1}·kg^{-1}: (3.75 × 1000)/110 kg = 34.1

34.1 = 3.5 + 2(workload)/110

30.6 = 2(workload)/110

30.6/2 = workload/110

15.3 = workload/110

15.3 × 110 = workload

1683 = workload

Resistance setting:

1683 = (resistance setting)(80)(6)

1683 = (resistance setting)(480)

1683/480 = resistance setting

3.506 = resistance setting

26. Answer: 5.6 kg

Solution:

12 METs × 3.5 = 42 ml·min^{-1}·kg^{-1}

42 = 3.5 + 2(workload)/70

38.5 = 2(workload)/70

38.5/2 = workload/70

19.25 = workload/70

19.25 × 70 = workload

1347.5 kg·m·min^{-1} = workload

Resistance setting:

1347.5 = (resistance setting)(3)(80)

1347.5 = (resistance setting)(240)

1347.5/240 = resistance setting

5.614 = resistance setting

27. Answer: 30.2 ml·min^{-1}·kg^{-1}

Solution:

VO$_2$ = 3.5 + 3(800)/90

$$= 3.5 + 26.7$$

$$= 30.2$$

28. Answer: 1.2 L·min^{-1}

Solution:

50 watts × 6 = 300 kg·m·min^{-1}

VO_2 = 3.5 + 3(300)/90

$\quad\quad$ = 3.5 + 10

$\quad\quad$ = 13.5 ml·min^{-1}·kg^{-1}

Convert to L·min^{-1}: (13.5 × 90 kg)/1000 = 1.215

29. Answer: $7.1 \text{ ml·min}^{-1}\text{·kg}^{-1}$

Solution:

Workload = (0.5 kg)(2.4 m·rev^{-1})(50) = 60 kg·m·min^{-1}

VO_2 = 3.5 + 3(60)/50

$\quad\quad$ = 3.5 + 3.6

$\quad\quad$ = 7.1

30. Answer: 5.2 hours

Solution:

VO_2 = 3.5 + 2(1000)/70

$\quad\quad$ = 3.5 + 28.6

$\quad\quad$ = 32.1 ml·min^{-1}·kg^{-1}

Convert to L·min^{-1}: (32.1 × 70)/1000 = 2.247 L·min^{-1}

2.247 × 5 kcal·L^{-1} = 11.2 kcal·min^{-1}

3500 kcal/11.2 = 312.5 minutes

312.5/60 = 5.2

31. Answer: $21.6 \text{ ml·min}^{-1}\text{·kg}^{-1}$

Solution:

VO_2 = 0.35(20) + 0.061(20)(12)

$\quad\quad$ = 7 + 14.64

$\quad\quad$ = 21.64

32. Answer: 4.95 L·min^{-1}

Solution:

VO_2 = 0.35(35) + 0.061(35)(20)

$\quad\quad$ = 12.25 + 42.7

$\quad\quad$ = 54.95 ml·min^{-1}·kg

Convert to L·min^{-1}: (54.95 × 90 kg)/1000 = 4.9455

33. Answer: 13.4 kcal·min^{-1}

    Solution:

    $VO_2 = 0.35(40) + 0.061(40)(10)$

    $= 14 + 24.4$

    $= 38.4$ ml·min^{-1}·kg^{-1}

    Convert to L·min^{-1}: $(38.4 \times 70$ kg$)/1000 = 2.688$

    2.688×5 kcal·min$^{-1} = 13.4$

34. Answer: 9.6 inches or approximately 3 platforms

    Solution:

    $VO_2 = 11$ METs $\times 0.85 \times 3.5 = 32.7$ ml·min^{-1}·kg^{-1}

    $32.7 = 0.35(35) + 0.061(35)$(height)

    $32.7 = 12.25 + 2.135$(height)

    $20.45 = 2.135$(height)

    $20.45/2.135 = $ height

    $9.58 = $ height

35. Answer: 56 beats·min^{-1}

    Solution:

    Convert L·min^{-1} to ml·min^{-1}·kg^{-1}:

    $(1.6 \times 1000)/90 = 17.8$

    $17.8 = 0.35$(rate) $+ 0.061$(rate)(15)

    $17.8 = 0.35$(rate) $+ 0.915$(rate)

    $17.8 = 1.265$(rate)

    $17.8/1.265 = $ rate

    14 steps·minute$^{-1} = $ rate

    Set the metronome at $4 \times 14 = 56$

36. Answer: 28.0 ml·min^{-1}·kg^{-1}

    Solution:

    $VO_2 = 3.5(8$ METs$)$

    $= 28.0$

37. Answer: 10

    Solution:

    Calculate desired VO_2: $0.70 \times 50 = 35$ ml·min^{-1}·kg^{-1}

    $35 = 3.5$(MET level)

    $35/3.5 = $ MET level

    $10 = $ MET level

38. Answer: 307 kcal

    Solution:

    Convert watts to kg·m·min^{-1}: $150 \times 6 = 900$ kg·m·min^{-1}

$VO_2 = 3.5 + 2(900)/70$

$= 3.5 + 1800/70$

$= 3.5 + 25.71$

$= 29.21$

Convert to $L \cdot min^{-1}$: $(29.21 \times 70)/1000 = 2.045\ L \cdot min^{-1}$

Convert to $kcal \cdot min^{-1}$: $2.045 \times 5 = 10.225\ kcal \cdot min^{-1}$

10.225×30 minutes $= 306.75$

39. Answer: Level 1.9

Solution:

Calculate desired VO_2: $0.75 \times 30 = 22.5\ ml \cdot min^{-1} \cdot kg^{-1}$

$22.5 = 3.5 + 3(workload)/90$

$22.5 - 3.5 = 3(workload)/90$

$19 = 3(workload)/90$

$(19 \times 90)/3 = workload$

$570\ kg \cdot m \cdot min^{-1} = workload$

Divide by 300 to get Airdyne level:

$570/300 = 1.9$

40. Answer: 15 mph

Solution:

Convert METs to $ml \cdot min^{-1} \cdot kg^{-1}$: $8 \times 3.5 = 28\ ml \cdot min^{-1} \cdot kg^{-1}$

$28 = 3.5 + (tabled\ value)/60$

$28 - 3.5 = (tabled\ value)/60$

$24.5 \times 60 = tabled\ value$

$1470 = tabled\ value$

Our small cyclist is looking for a net VO_2 of $1470\ ml \cdot min^{-1}$. According to Table 5.1, a speed of 15 mph yields a net VO_2 of $1510\ ml \cdot min^{-1}$; close enough.

41. Answer: 160 bpm

Solution:

target HR $= 0.75[(220 - 30) - 70] + 70$

$= 0.75(190 - 70) + 70$

$= 0.75(120) + 70$

$= 90 + 70$

$= 160$

42. Answer: 99 bpm

Solution:

target HR $= 0.45(120 - 82) + 82$

$= 0.45(38) + 82$

$= 17.1 + 82$

$= 99.1$

43. Answer: 179 bpm

    Solution:

    target HR = 0.85[(220 − 20) − 60] + 60

                = 0.85(220 − 60) + 60

                = 0.85(140) + 60

                = 119 + 60

                = 179

44. Answer: 123 bpm

    Solution:

    target HR = 0.82(220 − 70)

                = 0.82(150)

                = 123

45. Answer: 176

    Solution:

    target HR = 0.89(220 − 22)

                = 0.89(198)

                = 176.22

    The target HR is 176 bpm. The number of beats that occurs in 10 seconds is one sixth this value: 176/6 = 29.34. Alternatively, you can look at Table 6.2 and find the 10-second count that most closely corresponds to the target HR.

46. Answer: 2.24 L·min⁻¹

    Solution:

    Look it up in Table 7.1: 32 ml·min⁻¹·kg⁻¹

    Convert to L·min⁻¹: (32 × 70)/1000 = 2.24

47. Answer: 30 ml·min⁻¹·kg⁻¹

    Solution:

    VO_2max = [10.51(195) + 6.35(90) − 10.49(45) + 519.3]/90

                = [2049.45 + 571.5 − 472.05 + 519.3]/90

                = 2668.2/90

                = 29.647

48. Answer: 23 ml·min⁻¹·kg⁻¹

    Solution:

    VO_2max = [6965.2 + 20.02(75) − 25.7(52) + 595.5(1) − 224(20.28) − 11.5(130)]/75

                = [6965.2 + 1501.5 − 1336.4 + 595.5 − 4542.72 − 1495]/75

                = 1688.08/75

                = 22.508

49. Answer: 28 ml·min^{-1}·kg^{-1}

Solution:

From the Astrand nomogram, the raw VO$_2$max is approximately 2.4 L·min^{-1}, and the age correction factor is 0.83. Thus, the absolute VO$_2$max is 2.4 × 0.83 = 1.992; the relative VO$_2$max is (1.992 × 1000)/70 = 28.457 ml·min^{-1}·kg^{-1}.

50. Answer: 21.3 ml·min^{-1}·kg^{-1}

Solution:

The age-predicted maximal HR is 220 − 30 = 190 bpm. By plotting HRs on the blank graph provided at the end of Chapter 7, the estimated maximal workload at that HR is 800 kg·m·min^{-1}. Now we enter that into the cycle ergometry equation and solve for VO$_2$.

$$VO_2 = 3.5 + 2(800)/90$$
$$= 3.5 + 1600/90$$
$$= 21.278$$

Multiple Choice Test

Note: Questions marked by an asterisk are based on supplemental material that goes beyond the scope of the American College of Sports Medicine guidelines.

QUESTIONS

1. What would be the energy expenditure for a 73-kg woman walking at 3.9 mph and 7% grade?

    a. 7.7 METs

    b. 27 ml·min^{-1}·kg^{-1}

    c. 2.7 L·min^{-1}

    d. Two of the above

2. Which of the following would require your 65-kg client to exercise at 4.1 METs?

    a. Walking at 4.0 mph

    b. Running at 7.0 mph

    c. Cycling at 1200 kg·m·min^{-1}

    d. None of the above

3. Your client weighs 250 lb and is walking on a treadmill at 2.8 mph and 4% grade, 5 days a week, for 60 minutes each exercise session. How many weeks must your client exercise to lose 30 lb of fat (assume caloric intake is constant)?

    a. 52

    b. 48

    c. 23

    d. 38

4. A peripheral vascular disease patient cannot exceed 2.4 mph on the treadmill but needs to exercise at 6.0 METs. What percentage grade do you set?

    a. 9.6

    b. 11.2

    c. 5.8

    d. 0

5. A patient needs to exercise at 5 METs while walking. Assuming a 7% grade, what speed would the patient walk at?

    a. 3.3

    b. 3.8

    c. 2.3

    d. 4.0

6. Which of the following would provide a metabolic cost equal to 17.6 METs for a 62-kg client?

    a. Running on a treadmill at 6 mph and 18% grade

    b. Cycling at 1200 kg·m·min^{-1}

    c. Stepping at a rate of 50 steps per minute on a 12-inch platform

    d. Walking at 4 mph and 5% grade

7. Your client weight 200 lb. You want him to burn 0.5 lb of fat each week from exercise. He plans to run outdoors 3 times a week, at 5 mph. How long should he run each day?

    a. 49 minutes

    b. 40 minutes

    c. 42 minutes

    d. 34 minutes

8. Your client, a 70-kg woman, has finished a 5-mile race in 33 minutes, 24 seconds. What was her VO$_2$ during the race?

    a. 36 ml·min^{-1}·kg^{-1}

    b. 5.6 METs

    c. 5.2 L·min^{-1}

    d. 3.6 L·min^{-1}

9. Consider the previous question. If your client was climbing a 4% grade outdoors, by how much would she have to slow her speed to maintain the same VO_2?

    a. 8.9 mph

    b. 2.4 mph

    c. 6.5 mph

    d. 0 mph

10. A 70-kg client has a VO_2 of 2.6 $L \cdot min^{-1}$. Which activity and intensity would elicit this VO_2?

    a. Running 10 mph

    b. Exercising on a stationary cycle at 1200 $kg \cdot m \cdot min^{-1}$

    c. Stepping at a rate of 29 steps per minute up a 9-inch platform

    d. Running outdoors at 10 mph 6% grade

11. What would the previous client expend in kilocalories if he exercised at the above intensity for 45 minutes?

    a. 130

    b. 585

    c. 390

    d. 620

12. Your 70-kg client wants to exercise at 10 METs on a Tunturi bike. If she pedals at 60 rpm, what resistance setting in kg is needed?

    a. 5.0

    b. 7.2

    c. 8.4

    d. 6.1

*13. An 85-kg individual is exercising at a VO_2 of 35 $ml \cdot min^{-1} \cdot kg^{-1}$. What type and intensity of exercise might he be doing?

    a. Schwinn Airdyne at a setting of 4

    b. Arm cranking a Monark leg ergometer at 900 $kg \cdot m \cdot min^{-1}$

    c. Tunturi bike at 60 rpm and 1.2 kg

    d. Rowing at 200 watts

14. You want your 95-kg client to burn 3500 kcal per week on a stationary bike. If he exercises at 1200 $kg \cdot m \cdot min^{-1}$, how many minutes must he exercise?

    a. 257

    b. 275

    c. 422

    d. 294

15. Your 85-kg friend is stepping up a 15-inch step at 30 steps per minute. What is her energy expenditure in $kcal \cdot min^{-1}$?

    a. 37.9

    b. 3.2

    c. 16.1

    d. 10.8

16. You teach a step aerobics class that uses 2.5-inch stepping platforms. The music you selected has a beat of 120 bpm. Assume that the average maximum MET level for the participants in your class is 10 METs. How many step platforms should be used if you want your class to exercise at 70% of their capacity?

    a. 3

    b. 2

    c. 1

    d. None of the above

*17. Exercising at 120 watts on a Concept II rower for 30 minutes burns how many kilocalories for a 90-kg man?

    a. 280

    b. 363

    c. 263

    d. 632

*18. Which of the following will permit a 70-kg client to exercise at 70% of his 50 $ml \cdot min^{-1} \cdot kg^{-1}$ maximum?

    a. Running at 4.2 mph

    b. Walking at 2.0 mph

    c. Level 3.7 on an Airdyne cycle

    d. Rowing at an average power output of 100 watts

19. You have a 74-year-old relative who has a maximum HR of 130 bpm and a resting HR of 88. What would be his target HR at 50% of VO_2max using the percentage HRR method?

    a. 109 bpm

    b. 117 bpm

    c. 94 bpm

    d. 126 bpm

*20. During an aerobics class, how many heart beats in 10 seconds should your 29-year-old client obtain if he wants to maintain his workout at 70% of VO_2max? Use Table 6.1 and the percentage HRmax method.

    a. 133

    b. 157

    c. 26

    d. 50

*21. Your 95-kg client completes 6 minutes and 30 seconds on the Bruce protocol treadmill test. What is the VO_2max in $L \cdot min^{-1}$?

    a. 4.0

    b. 2.1

    c. 3.4

    d. 1.2

22. Your 40-year-old, 70-kg client performs a submaximal bike test. His HRs at the end of each stage are 112 bpm at 150 $kg \cdot m \cdot min^{-1}$, 120 at 300 $kg \cdot m \cdot min^{-1}$, and 140 at 450 $kg \cdot m \cdot min^{-1}$. What is his VO_2max in $ml \cdot min^{-1} \cdot kg^{-1}$?

    a. 40

    b. 35

    c. 18

    d. 29

23. What would the MET level for an 85-kg person walking on a treadmill at 2.7 mph and 9% grade be?

    a. 9.3

    b. 6.4

    c. 8.2

    d. 22.3

24. What would be the VO_2 in $ml \cdot min^{-1} \cdot kg^{-1}$ for a 100-kg man to walk at 2.7 mph and 8% grade?

    a. 30

    b. 6.0

    c. 21

    d. 42

25. A 62-kg client walks on a treadmill at 3.2 mph and 12% grade. What is the metabolic cost?

    a. 27.0 $ml \cdot min^{-1} \cdot kg^{-1}$

    b. 3.05 $L \cdot min^{-1}$

    c. 12.5 kcal/min

    d. 30.5 $ml \cdot min^{-1} \cdot kg^{-1}$

26. Calculate the VO_2 in $L \cdot min^{-1}$ for a 90-kg client who runs on a treadmill at 9 mph up a 5% incline.

    a. 62

    b. 17.7

    c. 3

    d. 5.6

27. Running outdoors at 7 mph elicits what VO_2 in $ml \cdot min^{-1} \cdot kg^{-1}$?

    a. 41

    b. 56

    c. 29

    d. 85

*28. Which of the following would allow a 90-kg man to expend 36 $kcal \cdot min^{-1}$?

    a. Walking 4.0 mph up a 9% grade

    b. Running on a treadmill at 10 mph and 10% grade

    c. Jogging at 7.0 mph and 0% grade

    d. Rowing at 150 watts

29. A 100-kg friend is exercising on a stationary bike at 300 watts. What is his estimated VO_2?

    a. 4.0 METs

    b. 5.2 $L \cdot min^{-1}$

    c. 39.5 $ml \cdot min^{-1} \cdot kg^{-1}$

    d. 52.1 $ml \cdot min^{-1} \cdot kg^{-1}$

30. You want your 90-kg client to exercise at 75% of his 3.5 L·min^{-1} capacity on a Monark bike. If he pedals at 70 rpm, what resistance setting in kg is needed?

    a. 2.7

    b. 3.0

    c. 1.0

    d. 4.5

31. A 70-kg patient with cardiac disease is arm cranking on a Monark arm ergometer at 50 rpm with a resistance setting of 1.0 kg. What is the VO$_2$ in ml·min^{-1}·kg^{-1}?

    a. 4.6

    b. 8.6

    c. 27.2

    d. 12.2

32. What is the metabolic cost for a 70-kg client while stepping on a 14-inch step at a rate of 15 steps per minute?

    a. 5.2 METs

    b. 20 ml·min^{-1}·kg^{-1}

    c. 7.0 METs

    d. 2.8 L·min^{-1}

33. You want your 250-lb client to exercise at 2.5 L·min^{-1} while stepping on a 12-inch bench. At what rate would you set the metronome?

    a. 80

    b. 120

    c. 60

    d. 100

*34. Which of the following would allow your 51-kg client to exercise at 70% of his 40 ml·min^{-1}·kg^{-1} capacity?

    a. Level 8 on the Stairmaster

    b. Running at 7.2 mph

    c. Cycling at 1.4 kg and 60 rpm on a Monark bike

    d. Stepping at 40 steps·min^{-1} on a 14-inch bench

35. Your 27-year-old client wants to exercise at 80% of her functional capacity. Her resting HR is 60 bpm. What is her target HR using the percentage HRR method?

    a. 133 bpm

    b. 106 bpm

    c. 166 bpm

    d. 154 bpm

36. Your 19-year-old friend has a resting HR of 72 bpm. What would be her target HR at 90% of VO$_2$max using the percentage HRR method?

    a. 162 bpm

    b. 180 bpm

    c. 172 bpm

    d. 188 bpm

*37. A 55-year-old, 72-kg man completes the 255-watts stage of the Storer cycling protocol. What is his VO$_2$max in ml·min^{-1}·kg^{-1}?

    a. 42.8

    b. 36.8

    c. 29.1

    d. 18.3

38. Your 25-year-old, 95-kg male client performs an Astrand bike test. His average HR for the fifth and sixth minute at 900 kg·m·min^{-1} is 150 bpm. What is his VO$_2$max in ml·min^{-1}·kg^{-1}?

    a. 18.7

    b. 26.3

    c. 32.6

    d. None of the above

39. What would be the energy expenditure in kcal for an 80-kg person who is walking 3.0 mph on a treadmill up a 12% grade for 60 minutes?

    a. 550

    b. 3520

    c. 692

    d. 11.5

40. Your 50-kg client is walking on the treadmill at 2.0 mph, 15% grade. What is the

total oxygen consumption in liters for a 30-minute exercise session?

    a. 24.8

    b. 34.9

    c. 16.9

    d. 1.2

41. Running on a treadmill to completion of stage 4 of the Bruce protocol requires what MET level?

    a. 42.1

    b. 50.3

    c. 9

    d. 12

42. Your running partner has a functional capacity of 15 METs. You want him to train at 80% of his functional capacity. He will be running on a treadmill at 6 mph. What percentage grade should he use?

    a. 7.3

    b. 2.9

    c. 5.0

    d. 4.4

43. You have a VO_2max of 70 $ml \cdot min^{-1} \cdot kg^{-1}$ and want to exercise at 85% of it on a treadmill. If the speed is 7.0 mph, at what should the grade be set?

    a. 11

    b. 12

    c. 10

    d. 9

44. What workload in $kg \cdot m \cdot min^{-1}$ should be prescribed if you want your 90-kg client to exercise at 10 $kcal \cdot min^{-1}$ on a stationary bike?

    a. 900

    b. 842

    c. 762

    d. 1200

45. An 85-kg client is arm cranking a Monark leg ergometer at 100 watts. What is the VO_2 in $L \cdot min^{-1}$?

    a. 2.1

    b. 3.0

    c. 1.5

    d. 1.2

46. What is the metabolic cost for your 90-kg client who uses a 10-inch step at a rate of 25 steps per minute?

    a. 6.1 METs

    b. 2.4 $L \cdot min^{-1}$

    c. 10.8 kcal/min

    d. None of the above

*47. Your 88-kg client is exercising at level 10 on a Stairmaster. What is the energy expenditure?

    a. 10 METs

    b. 25 $ml \cdot min^{-1} \cdot kg^{-1}$

    c. 17.5 kcal/min

    d. 3.4 $L \cdot min^{-1}$

*48. Your 70-kg friend wants to ride his bicycle outdoors at 9 METs. At what speed should he ride?

    a. 10 mph

    b. 11 mph

    c. 15 mph

    d. 18 mph

*49. You want your 35-year-old client to exercise at 60% of VO_2max. Using Table 6.1 and the percentage HRmax method, what would be his target HR?

    a. 180 bpm

    b. 146 bpm

    c. 141 bpm

    d. 173 bpm

*50. Your 30-year-old, 90-kg male client has an estimated VO_2max of 20 $ml \cdot min^{-1} \cdot kg^{-1}$. Which of the following sets of test data correspond to this value?

    a. He has completed a 345-watts stage of the Storer cycling protocol.

    b. He has completed a 240-watts stage of the Storer cycling protocol.

    c. He has completed a 1-mile walk in 23 minutes and 15 seconds, with a post-walk HR of 135 bpm.

    d. He has completed a 1-mile walk in 20 minutes and 27 seconds, with a post-walk HR of 160 bpm.

Answers

1. b	14. a	27. a	40. b
2. a	15. c	28. b	41. d
3. d	16. a	29. c	42. d
4. a	17. c	30. a	43. a
5. c	18. c	31. b	44. b
6. a	19. a	32. a	45. a
7. c	20. c	33. a	46. c
8. d	21. b	34. a	47. a
9. b	22. d	35. c	48. d
10. b	23. b	36. d	49. c
11. b	24. c	37. a	50. c
12. d	25. d	38. c	
13. b	26. d	39. c	